SpringerBriefs in Computer Science

Series Editor

Stan Zdonik
Peng Ning
Shashi Shekhar
Jonathan Katz
Xindong Wu
Lakhmi C Jain
David Padua
Xuemin (Sherman) Shen
Borko Furht
V. S. Subrahmanian
Martial Hebert
Katsushi Ikeuchi
Bruno Siciliano

For further volumes:
http://www.springer.com/series/10028

Hebert Montegranario • Jairo Espinosa

Variational Regularization of 3D Data

Experiments with MATLAB®

 Springer

Hebert Montegranario
Mathematics Institute
Universidad de Antioquia
Medellín
Colombia

Jairo Espinosa
Universidad Nacional de Colombia
Medellín
Colombia

MATLAB® is a registered trademark of The MathWorks, Inc.

References to MATLAB® and other copyrighted trademarks, servicemarks, marks and registered marks owned by The MathWorks, Inc. and/or its susidiaries may appear in this book. Rather than use a trademark symbol with every occurrence of a trademarked name, logo, or image, we use the names, logos, and images only in an editorial fashion with no intention of infringement of the trademark.

For MATLAB® and Simulink® product information, please contact:
The MathWorks, Inc.
3 Apple Hill Drive
Natick, MA, 01760-2098 USA
Tel: 508-647-7000
Fax: 508-647-7001
E-mail: info@mathworks.com
Web: mathworks.com

ISSN 2191-5768   ISSN 2191-5776 (electronic)
ISBN 978-1-4939-0532-4   ISBN 978-1-4939-0533-1 (eBook)
DOI 10.1007/978-1-4939-0533-1
Springer New York Heidelberg Dordrecht London

Library of Congress Control Number: 2014931082

Printed on acid-free paper

Springer is part of Springer Science+Business Media (www.springer.com)

For Hanna and Santiago
For Laura and Jairo

Preface

The overwhelming amount of applications and information produced by modern technology is supported on subtle mathematical methods. Variational method is one of them. More than a technique, it has developed to become a very general point of view that solve a class of problems, too wide to be summarized in just one book.

One of the first applications of variational ideas in data modelling was the creation of variational splines theory, introduced in the mathematical literature by I. J. Schoenberg in 1946. Since then, splines have been applied in many branches of mathematics, such as approximation theory, numerical treatment of differential and integral equations, and statistics. In addition, they have become useful tools in many field of applications, especially CAGD, manufacturing, animation and tomography. Currently there exists a great amount of research in these fields, thus there is a great demand for a text of this kind that offers an introduction to the theme from the very beginning of variational calculus and that can be used to understand more advanced material.

In the area of surface reconstruction the goal is to obtain a digital representation of a real, physical object or phenomenon described by a set of points, which is sampled on or near the object's surface. This representation is a continuous model obtained from a discrete number of samples by interpolation or approximation. Recently there has been a growing interest in this field motivated by the increased availability of point-cloud data obtained from medical scanners, laser scanners, vision techniques (e.g. range images), and other modalities.

This book is an introduction to variational methods for data modelling and its application in computer vision. We see interpolation as an inverse problem that can be solved by Tikhonov regularization. The solutions are generalizations of one-dimensional splines, applicable to n-dimensional data. The central idea is that these splines can be obtained by regularization theory, using a trade-off between fidelity to data and smoothness properties; as a consequence, they are applicable both in interpolation and approximation of exact or noisy data.

In order to obtain a self-contained text, we provide the necessary fundamentals of functional analysis and variational calculus as well as splines. The implementation and numerical experiments are illustrated using Matlab. The discussion includes the necessary theoretical background for approximation methods and some details of the computer implementation of the algorithms. A working knowledge of multi-variable calculus and basic vector and matrix methods should serve as an adequate prerequisite.

Hebert Montegranario
Jairo Espinosa

Contents

Chapter 1
3D Data in Computer Vision and Technology

Three dimensional data appears in a wide range of modern technology applications, all of them are strongly influenced by the problems of computational vision. It is not strange because through vision, we obtain an understanding of what is in the world, where objects are located, and how they are changing with time. Human beings obtain this understanding as a gift from nature, with no effort, and without conscious introspection, so we may believe that vision is a very simple task to perform. Computational studies has helped us to understand the complexity of vision and the processes needed to perform visual tasks successfully.

In the second half of the twentieth century, there were two important attempts to provide a theoretical framework for understanding vision, by David Marr [43] and James Gibson [22]. Many applications of variational regularization are strongly influenced by Marr's **computational theory of vision.** Marr emphasized that vision was nothing more than an information-processing task. Any such task, he argued, could be described on three levels: (i) computational theory; (ii) specific algorithms; and (iii) physical implementation.

These levels can be considered independently, and this concept has become a paradigm of vision research. Marr proposed to create a computational theory for vision as a whole, arguing that visual processing passes through a series of stages, each corresponding to a different representation, from retinal image to 3D model representation of objects. Today, 40 years on, most would agree that Marr's framework for investigating human vision and, in particular, his strategy of dividing the problem into different levels of analysis, has become unquestioned.

The first part of vision -from images to surfaces- has been termed **early vision** and consists of a set of processes that recover physical properties of visible three-dimensional surfaces from the two dimensional images. Different arguments suggest that early vision processes correspond to conceptually independent modules that as first approximation can be studied in isolation. **Surface or 3D reconstruction** is one of these modules. Other early vision modules are edge detection, spatiotemporal interpolation and approximation, computational optical flow, shape from contours, shape from texture, shape from shading and binocular stereo.

Computer vision is an inter-disciplinary research field that has as primary objective to address visual perception through mathematical modelling of visual

H. Montegranario, J. Espinosa, *Variational Regularization of 3D Data,*
SpringerBriefs in Computer Science, DOI 10.1007/978-1-4939-0533-1_1,
© The Author(s) 2014

understanding tasks. It has evolved during the last decades with two main goals: to develop image understanding systems and to understand biological vision. Its main focus is on theoretical studies of vision, considered as an information processing task.

Most early vision problems are ill-posed in the sense defined by Hadamard [28]. A problem is well posed when its solution (a) *exists*, (b) *is unique*, and (c) *depends continuously on the initial data*. Ill-posed problems fail to satisfy at least one of these criteria. Authors like Poggio [52] show precisely the mathematical structure of these problems. This fact suggests the use of **regularization methods** developed in mathematical physics for solving the ill-posed problems of early vision.

The main idea supporting **Tikhonov regularization** theory [63] is that the solution of an ill-posed problem can be obtained using a variational principle, which contains both the data and prior smoothness information. Considering, for instance, 3D data, $\{(x_i, y_i, z_i)\}_{i=1}^M$ these two features are taken into account, assuming $a_i = (x_i, y_i)$, $z_i = f(a_i)$ and minimizing a functional of the form

$$J[f] = \sum_{i=1}^M (f(a_i) - z_i)^2 + \lambda R[f].$$

With this approach, we are looking for an approximation that is simultaneously smooth and close to the data. Smoothness is included with the smoothness functional or **regularizer** $R[f]$ in such a way that lower values of the functional corresponds to smoother functions; λ is a positive number called **regularization parameter**.

In the applications of surface reconstruction the goal is to obtain a digital representation of a physical object or phenomenon described by a set of points, which are sampled on or near the object's surface. Recently there has been a growing interest in this field motivated by the increased availability of point-cloud data obtained from medical scanners, laser scanners, vision techniques (e.g. range images), and other modalities.

Apart from being ill-posed, the problem of surface reconstruction from unorganized point clouds is challenging because the topology of the real surface can be very complex, and the acquired data may be non-uniformly sampled and contaminated by noise. Moreover, the quality and accuracy of the data sets depend upon the methodologies which have been employed for acquisition (i.e. laser scanners versus stereo using uncalibrated cameras). Furthermore, reconstructing surfaces from large datasets can be prohibitively expensive in terms of computations. Our approach to surface reconstruction will be based in the following multivariate interpolation problem:

Given a discrete set of scattered points $\mathcal{A} = \{a_1, a_2, \ldots a_M\} \subset \mathbb{R}^n$ and a set of possible noisy measurements $\{z_i\}_{i=1}^M$, find a continuous or sufficiently differentiable function

$$f : \mathbb{R}^n \to \mathbb{R}$$

such that f interpolates ($f(a_i) = z_i$) or approximates ($f(a_i) \approx z_i$) the data

1.1 Applications of 3D Reconstruction

Surface reconstruction is an important problem in computational geometry, computer aided design (CAD), computer vision, graphics, and engineering. The problem of building surfaces from unorganized sets of 3D points has recently gained a great deal of attention. In fact, in addition to being an interesting problem of topology extraction from geometric information, its applications are becoming more and more numerous. For example, the acquisition of large numbers of 3D points is becoming easier and more affordable using, for example, 3D-scanners [53].

Range images are a special class of digital images. Each pixel of a range image expresses the distance between a known reference frame and a visible point in the scene, therefore, a range image reproduces the 3D structure of a scene. Range images are also referred to as depth images, xyz maps, surface profiles and 2.5D images. They can be represented in two basic forms. One is a list $\mathcal{A} = \{a_1, a_2, \ldots a_M\} \subset \mathbb{R}^3$ of M scattered points in 3D coordinates $a_i = (x_i, y_i, z_i)$ in a given reference frame (point clouds), without any particular order. The other is given as a matrix of depth values of points along the directions of the x, y image axes, which makes spatial organization explicit. Range images represent the position of surface more directly, intensity images are of limited use in terms of estimation of surfaces.

In many applications, objects are better described by their external surface rather than by the mere set of unorganized data (clouds of points, data slices, etc.). For example, in medical applications based on CAT scans or NMRs it is often necessary to visualize some specific tissues such as the external surface of an organ starting from the acquired 3D points. This can be achieved by selecting the points that belong to a specific class (organ boundary, tissue, etc.) and then generating the surface from their interpolation.

Three dimensional object surface reconstruction and modeling plays a very important role in **reverse engineering** [67]. For example, if we want to build a geometrical model to a 3D object for reproduction, an efficient way is to scan the object with a digital data-acquisition device and perform surface reconstruction to get its geometrical model. Obtaining a computer model provides several advantages in improving the quality and efficiency of design, manufacture and analysis. Reverse engineering typically starts with measuring an existing object so that a surface or solid model can be deduced in order to apply CGD/CAM technologies.

There are several application areas of reverse engineering. For example, in the automobile industry it can be applied to produce a copy of a part, when no original drawings or documentation are available. In other cases we may want to re-engineer an existing part, when analysis and modifications are required to construct a new improved product. Another important area of application is to fit some surfaces of human body, for mating parts such as prostheses or space suits.

Three dimensional surface points can be obtained by tactile methods such as Coordinate Measuring Machines (CMM), or using non-contact methods, such as magnetic field measurement machines and optical range scanners. After surface point acquisition, they must be translated to a common coordinate system and merge the overlapping points of neighbouring data sets. Finally comes the process of surface reconstruction (surface meshing/triangulation) and rendering.

From the point of view of technology, 3D reconstruction methods can be collected under two groups: active and passive. Active methods make use of calibrated light sources such as lasers or coded light most typical example of which is the shape from optical triangulation method. Passive methods on the other hand, extract surface information by the use of 2D images of the scene. Among the most common that fall into this category are the techniques known as shape from silhouette, shape from stereo, and shape from shading.

There are several techniques that recover shape from some kind of data and they are all inverse problems commonly called shape-from-X such that X could be texture, stereo, shading or other. For example the shape from shading problem is to compute the three-dimensional shape of a surface from the brightness of one black and white image of that surface.

Shape from texture is a computer vision technique where a 3D object is reconstructed from a 2D image. The creation of a system able to simulate that behaviour is very complex, nevertheless, human vision is capable to realize patterns, estimate depth and recognize objects in an image by using texture as a stimulus. Some of these problems, as for example, shape from shading can be treated by variational methods.

Next we give a brief description of this books contents. In the next chapter we provide the general framework of function spaces with norms and inner product, an indispensable tool for dealing with the problem of reconstruction from data in one or more dimensions. These ideas from functional analysis have shown to be very useful in modern applications. Chapter 3 contains an introduction to variational methods and its classical techniques. This is the core theory that supports the basic understanding of regularization methods and is an introduction to variational theory of splines. In Chap. 4, the reader can understand in more detail the relation between the abstract setting given in former chapters and its application to concrete problems, splines arise as a solution to limitations of classical Lagrange interpolation and are generalized from one to several variables by extending the variational definition of splines to further dimensions.

Thanks to its formulation as minimization of functionals, this theory not only has a solid mathematical formulation but also a real interpretations; that is the subject of Chap. 5, where we provide some physical and geometrical metaphors that improve our comprehension of the problem. Chapter 6 gives us a bridge between reconstruction problems and the standard theory of inverse problems from the point of view of operator theory and finally, the last two chapters deal with the problem of 3D reconstruction from scattered data. The last chapter can also be used as an introduction to modern meshless methods. In recent years these methods have increased in number of applications. The Matlab code may help those who are looking for an introduction to these techniques.

Chapter 2
Function Spaces and Reconstruction

A reconstruction problem is solved when we are able to find a function that models
or describes the behavior of data. Although in each problem there will be a particular method to obtain or define this function, many of them are included in the theory
of function spaces. In this chapter we will see the alphabet that permit us to understand the language for the rest of the book. The basic idea is to consider functions
as simple points that behaves with the properties of Euclidean space, so we begin
by generalizing the ideas of distance and orthogonality of three dimensional space
to norms and inner products in function spaces. Next we study integration by parts
and its application to define distributions and Sobolev spaces. Distribution theory
builds a bridge between discrete and continuous processes by extending the concept
of differentiability of a function.

2.1   3D Reconstruction

In this book, our main goal is the reconstruction of the form of an object given a set
of three dimensional data sampled from a real object. In general we are dealing with
the problem of the complete reconstruction of a multivariate function $f : \mathbb{R}^n \to \mathbb{R}$
from sampling values $\{f(x_i)\}_{i=1}^{N}$.

  For the one-variable case a possible answer to this problem is the well-known
Shannon sampling theorem [34]. A band limited function f with frequency spec-
trum limited by the Nyquist frequency, can be reconstructed perfectly from its
regularly-spaced (ideal) samples:

Theorem1 (Shannon) Suppose $f \in \mathcal{L}^2$ and $\hat{f}(\xi) = 0$ for $|\xi| \geq \Omega$. Then f is com-
pletely determined by its values at the points $t_k = k\pi / \Omega$, $k = 0, \pm 1, \pm 2, \ldots$ by means
of

$$f(t) = \sum_{k=-\infty}^{\infty} f\left(\frac{k\pi}{\Omega}\right) \frac{\sin(\Omega t - k\pi)}{\Omega t - k\pi}.$$

This expression shows that $f(t)$ can be written as translates or convolution with
a single function. In the next chapters, we will show that the same situation may
happen for functions on $\mathbb{R}^n$. In these cases, it is necessary to deal with multivariable
functions and non-uniform sampling in which the location of measurement points

H. Montegranario, J. Espinosa, *Variational Regularization of 3D Data,*   5
SpringerBriefs in Computer Science, DOI 10.1007/978-1-4939-0533-1_2,
© The Author(s) 2014

may be irregular, either because it is not possible to control the measurement process or because some domain needs a particular emphasis. During the last years, this problem has been also widely studied and there exist a good number of results that share some of the features of Shannon's sampling theorem.

This can be generalized by changing the sinc function to an appropriate generating function $\phi(x)$ that results in spaces of the form $\left\{ \sum_{k \in \mathbb{Z}} c(k)\phi(x-k) \,|\, c \in \ell^2 \right\}$, where $\phi(x)$ does not have to be band-limited [66]. Extensions to the multidimensional irregular sampling of band-limited functions in very general spaces can be found in [34].

Our point of view to the problem will be to consider 3D reconstruction as an inverse problem that can be solved by variational regularization. Well-posedness of an inverse problem depends on the topological properties, therefore it is very important the choice of function spaces in which the optimization problems are going to be formulated and solved. We apply distribution theory as a fundamental tool for modeling and understanding reconstruction problems. Distributional spaces provides an abstract setting for including discrete and continuous function in the same framework. In this way it is possible to obtain explicit expressions for a family of interpolating and smoothing splines. This family includes the well-known cubic spline in one variable and the thin plate spline (TPS).

2.2 Function Spaces and Norms

Function spaces are vector spaces whose elements are functions. The collection of real valued functions u, v on a nonempty set Ω forms a real linear space (or vector space) with respect to the operations of pointwise addition: $(u+v)(x) = u(x) + v(x)$, $\forall x \in \Omega$ and scalar multiplication: $(\alpha u)(x) = \alpha u(x) \; \forall x \in \Omega \; \alpha \in \mathbb{R}$. If Ω is a nonempty open set in $\mathbb{R}^n$, an important example is $C^m(\Omega)$, the vector space of all real functions $f : \Omega \to \mathbb{R}$ that have continuous partial derivatives of orders $0, 1, \ldots, m$. If $f \in C^m(\Omega)$ for all $m = 0, 1, 2, \ldots$, then we write $f \in C^\infty(\Omega)$.

Definition 1 If $\mathcal{U}$ is a vector space a **norm** on $\mathcal{U}$ is a real valued function, denoted by $\|\cdot\|$, that satisfy three axioms

(i) $\|u\| > 0$ for each nonzero element u in $\mathcal{U}$.
(ii) $\|\alpha u\| = |\alpha| \|u\|$ for each $\alpha \in \mathbb{R}$ and each $u \in \mathcal{U}$.
(iii) $\|u+v\| \le \|u\| + \|v\|$ for all $u, v \in \mathcal{U}$. (Triangle Inequality)

A linear space $\mathcal{U}$ is a Banach space when is complete in the norm $\|\cdot\|$. The completeness condition means that every Cauchy sequence in the space converge to an element of the space.

Definition 2 We say a linear space $\mathcal{U}$ has an inner product if there is a symmetric bilinear form $B(u,v) = \langle u, v \rangle$, $\langle \cdot, \cdot \rangle : \mathcal{U} \times \mathcal{U} \to \mathbb{R}$ with the following properties for all u, v, w in $\mathcal{U}$

(i) $\langle u, v \rangle = \langle v, u \rangle$

(ii) $\langle u + v, w \rangle = \langle u.w \rangle + \langle v, w \rangle$

(iii) $\langle \alpha u, v \rangle = \alpha \langle v, u \rangle$

(iv) $\langle u, u \rangle \geq 0$ and $\langle u, u \rangle = 0$ if and only if $u = 0$ (positive definiteness)

Every inner product space $\mathcal{U}$ is a normed space with the norm $\| u \| = \sqrt{\langle u, u \rangle}$ of u. If $\mathcal{U}$ is a complete inner product space is called a *Hilbert Space*. As an important example, the space of real-valued continuous functions on $[0, 1]$ with the inner product

$$\langle u, v \rangle = \int_0^1 u(t)v(t)dt$$

is not complete (see [10] for a proof). Contrary, the set $\mathcal{L}^2[a, b]$ is a Hilbert space. This is the space of all complex-valued Lebesgue measurable functions on $[a, b]$ such that $\int_a^b |u(t)|^2 \, dt < \infty$. In $\mathcal{L}^2[a, b]$ the inner product is

$$\langle u, v \rangle = \int_a^b u(t)\overline{v(t)}dt$$

In an inner product space two vectors u, v are said to be orthogonal if $(u, v) = 0$. We symbolize this by $u \perp v$. A vector u is said to be orthogonal to a set $\mathcal{V}$ if $u \perp v$ for all $v \in \mathcal{V}$, and all the vectors u with this property form the orthogonal complement $\mathcal{V}^\perp$ of $\mathcal{V}$. The idea of orthogonality has many consequences in Hilbert spaces, similar to Euclidian geometry.

Definition 3 A real valued function $p(u) = |u|_{\mathcal{U}}$ defined on a linear space $\mathcal{U}$ is called a **seminorm** on $\mathcal{U}$, if the following conditions hold:

i. $p(u + v) \leq p(u) + p(v)$

ii. $p(\alpha u) = |\alpha| p(u)$

by this definition, every norm is a seminorm that can be seen as a norm with the requirement of positive definiteness removed. This condition is also removed from an inner product to obtain a **semi-inner product,** with corresponding seminorm $|u|_{\mathcal{U}}^2 = (u, u)$. This means that if $|u|_{\mathcal{U}} = 0$ it may happens $u \neq 0$. As a consequence the null space $\mathcal{N}$ of $|u|_{\mathcal{U}}$ is defined as

$$\mathcal{N} = \{u \in \mathcal{U} : |u|_{\mathcal{U}} = 0\}$$

Every vector space $\mathcal{U}$ with seminorm $|u|_{\mathcal{U}}$ induces a normed space $\mathcal{U}/\mathcal{N}$, called the quotient space. The induced norm on $\mathcal{U}/\mathcal{N}$ is clearly well-defined and is given by:

$$|u + \mathcal{N}|_{\mathcal{U}} = |u|_{\mathcal{U}}$$

Example 1 The thin plate energy

$$J[u] = \int_{\mathbb{R}^2} (u_{xx}^2 + 2u_{xy}^2 + u_{yy}^2)dydx$$

is an example of seminorm which is not a norm, its null space is $\mathcal{N} = \Pi_1(\mathbb{R}^2)$, the set of polynomials in two variables of degree less or equal than 1, thus $\Pi_1(\mathbb{R}^2) = span\{1, x, y\}$. As we will see in the next chapters, some variational problems of continuum mechanics and spline theory have to deal with optimization of seminorms.

2.3   Operators on Function Spaces

We say that two sets $\mathcal{U}$, $\mathcal{V}$ are connected by a functional dependency $A : \mathcal{U} \rightarrow \mathcal{V}$ if to each element $u \in \mathcal{U}$ there corresponds a unique element $v \in \mathcal{V}$. Roughly speaking, this functional dependency is called a *function* if the sets $\mathcal{U}$, $\mathcal{V}$ are sets of numbers; it is called a *functional* if $\mathcal{U}$ is a set of functions and $\mathcal{V}$ is a set of numbers, and it is called an *operator* if both sets are sets of functions [73].

Definition 4 An operator is a mapping $A : \mathcal{U} \rightarrow \mathcal{V}$ from one function space into another (or the same) function space. If $\mathcal{U}$ and $\mathcal{V}$ are linear spaces, then the mapping is **linear** if

$$A(u + v) = A(u) + A(v) \ and \ A(\alpha u) = \alpha A(u)$$

For all u, v in $\mathcal{U}$ and all real α. When an operator does not fulfill these conditions then it is called **nonlinear**

Example 2
1. Taking $\mathcal{U} = \mathcal{V} = \mathbb{R}^n$, any linear transformation represented by a $n \times n$ matrix is a linear operator
2. The Laplace operator $\Delta : C^2(\Omega) \rightarrow C(\Omega)$, $\Delta u = \dfrac{\partial^2 u}{\partial x^2} + \dfrac{\partial^2 u}{\partial y^2}$ with $\Omega \subset \mathbb{R}^2$, is a linear differential operator on the space of real valued functions with continuous second order derivatives on a region Ω in the plane into continuous functions in Ω.
3. For $\mathcal{U} = \mathcal{V} = C[0,1]$ and K continuous on $[0,1] \times [0,1]$ the mapping

$$A : C[0,1] \rightarrow C[0,1] \ with \ A(u)(t) = \int_0^1 K(\xi, t) u(\xi) d\xi$$

It is a linear integral operator.

Definition 5 If $\mathcal{U}$ and $\mathcal{V}$ are normed linear spaces, an operator $A : \mathcal{U} \rightarrow \mathcal{V}$ is said to be continuous at $u \in \mathcal{U}$ if for every sequence $\{u_k\} \subset \mathcal{U}$ converging to u

$$\lim_{k \rightarrow \infty} \| A(u_k) - A(u) \| = 0$$

Definition 6 A linear operator $A : \mathcal{U} \rightarrow \mathcal{V}$ is called bounded when there exists a real number B such that $\| A(u) \|_{\mathcal{V}} \leq B \| u \|_{\mathcal{U}}$ for all u in $\mathcal{U}$. The **norm** of a bounded linear operator is defined as $\|A\| = \sup\limits_{\|u\|=1} \|A(u)\|_V$. It can be shown that a linear transformation acting between normed linear spaces is continuous if and only if it is bounded [8].

Definition 7 The *null space* $\mathcal{N}(A)$ of a linear operator $A : \mathcal{U} \to V$ is the subspace of $\mathcal{U}$ given by

$$\mathcal{N}(A) = \{u \in \mathcal{U} : Au = 0\},$$

and the *range* $\mathcal{R}(A)$ of A is the subspace of $\mathcal{U}$ given by

$$\mathcal{R}(A) = \{v \in V : Au = v, \exists u \in \mathcal{U}\}$$

Definition 8 A real valued mapping $J : \mathcal{U} \to \mathbb{R}$ defined on a normed linear space $\mathcal{U}$ is called a **functional**. If the mapping is linear, it is called a **linear functional**. A linear functional can be seen as a special case of a linear operator with $V = \mathbb{R}$ then we can also speak about bounded functionals with norm $\|J\| = \sup\limits_{\|u\|=1} |J(u)|$.

Example 3
1. A class of very important functionals for modeling data are the evaluation functionals $L_{x_0}(f)$ defined on a function space $\mathcal{U}$. If $f \in \mathcal{U}$ $f : \Omega \to \mathbb{R}$ and $x_0 \in \Omega$, then $L_{x_0}(f) := f(x_0)$
   If $\mathcal{U}$ is a linear space then L_{x_0} is a linear functional since

$$L_{x_0}(\alpha u + v) = (\alpha u + v)(x_0) = \alpha u(x_0) + v(x_0) = \alpha L_{x_0}(u) + L_{x_0}(v)$$

2. If $\mathcal{U}$ is the linear space of integrable functions on $[a,b]$, then

$$J(u) = \int_a^b u(t)dt$$

   is a linear functional on $\mathcal{U}$
3. The functional

$$J(u) = \int_a^b \sqrt{1 + u'(t)^2}\, dt$$

   is a nonlinear functional defined in $C^1[a,b]$ that gives the arc length of the curve u in $[a,b]$.

The set of bounded linear functionals on $\mathcal{U}$ forms a linear space itself $\mathcal{U}'$ called the **dual space** of $\mathcal{U}$. By the Riesz representation theorem, on a Hilbert space the bounded linear functionals have a very simple form.

Theorem 2 (Riesz Representation Theorem) Every continuous linear functional $J : \mathcal{U} \to \mathbb{R}$ defined on a Hilbert space is of the form $J(u) = \langle u, R \rangle$ for an appropriate vector R that is uniquely determined by the given functional [8].

2.4 Distributions

Distribution theory was created mainly by Sobolev and Schwartz [19, 20, 32, 59] to give answers to problems of mathematical physics. However, as usual, after being rigorously formulated in mathematical terms, the theory has developed very far from its initial applications and became useful in other disciplines such as approximation theory. In the following chapters we will show how distribution spaces are especially suited for dealing with inverse problems of 3D reconstruction, providing a variational framework that conducts to a generalization from classical cubic spline to multivariate interpolation and approximation. The results of this approach include the well-known thin plate spline and other radial basis functions.

Although surface reconstruction is an ill-posed inverse problem, distribution theory gives a setting to construct spaces where they become well-posed. Discontinuous functions can be handled as easily as continuous or differentiable functions into a unified framework, making it appropriate for dealing with discrete data. We will show how this approach may serve as a tool for the double task of modelling these data while providing solutions for its reconstruction.

Distributions generalize functions by considering a function as a continuous linear functional on the space $C_0^\infty(\mathbb{R}^n)$ of infinitely differentiable functions with compact support. This setting is adequate for introducing the concept of generalized differentiation, which makes possible the calculus of distributions with all its practical consequences. The **delta functional** $\delta(x)$ contradictorily defined by Dirac as $\int \delta(x)\varphi(x)dx = \varphi(0)$; is a generalized function. The spaces of distributions have shown to be very useful for multivariate approximation and spline theory [37–39].

One important task is to find or construct spaces where approximation problems can be well-posed. Nevertheless, classical spaces like $C^m[a,b]$ may not be adequate for this purpose. For example, there may exist a sequence $\{f_k\}$ of functions in $C^1(\mathbb{R})$ such that they converge uniformly to f, but $f \notin C^1(\mathbb{R})$. Other example is the non invertibility of the order of differentiation, so it may happen that $\dfrac{\partial f}{\partial x \partial y} \neq \dfrac{\partial f}{\partial y \partial x}$. Distribution theory provides more versatile viewpoint for treating the problems attached to the operation of differentiation.

Definition 9 A multi-index α is an-tuple of nonnegative integers $\alpha = (\alpha_1, \alpha_2, ..., \alpha_n)$. The order of the multi-index α is the integer $|\alpha| = \sum_{k=1}^{n} \alpha_k$. Differential operators are defined using a multi-index $D^\alpha = \partial^\alpha = \dfrac{\partial^{|\alpha|}}{\partial x_1^{\alpha_1} \cdots \partial x_n^{\alpha_n}}$, on n real variables $x_1, x_2, ..., x_n$. For example if $n = 3$ and $\alpha = (1, 0, 2)$, then $D^\alpha u = \dfrac{\partial^3 u}{\partial x_1 \partial x_3^2}$.

The space $C^\infty(\Omega)$ consists of all functions $u : \Omega \subset \mathbb{R}^n \to \mathbb{R}$ such that $D^\alpha u \in C(\Omega)$ for each multi-index α. Then, u has continuous partial derivatives of all orders.

The space of **test functions** denoted $\mathcal{D}$, $\mathcal{D}(\Omega)$ or $C_0^\infty(\Omega)$ is formed by the elements φ of $C^\infty(\Omega)$ which have compact support. The support $Supp(\varphi)$ of φ is the closure of $\{x : \varphi(x) \neq 0\}$.

The set $C_0^\infty(\Omega)$ is dense in $\mathcal{L}^2(\Omega)$ the set of square integrable functions in the sense of Lebesgue. In other words, any function in $\mathcal{L}^2(\Omega)$, can be approximated by functions in $C_0^\infty(\Omega)$, in the sense that there is a sequence $\{u_k\}$ of functions in $C_0^\infty(\Omega)$, which converges to u in $\mathcal{L}^2(\Omega)$.

Every function $u \in C_0^k(\Omega)$ can be extended to a function of $C_0^k(\mathbb{R}^n)$, in this way, $C_0^k(\Omega)$ can be interpreted as a subspace of $C_0^k(\mathbb{R}^n)$. Thus, given an open set $\Omega \subset \mathbb{R}^n$, the set $C_0^k(\Omega)$ can be seen as the set of elements $u \in C_0^k(\mathbb{R}^n)$ for which $Supp(u) \subset \Omega$.

The space $\mathcal{D}$, of test functions introduces interesting properties in integration by parts formula becoming the basis for further developments in variational calculus [73].

The classic integration by parts formula with $-\infty < a < b < \infty$, states

$$\int_a^b u'\varphi dx = [u\varphi]_a^b - \int_a^b u\varphi' \, dx \text{ holds for } u, \varphi \in C^1[a,b] \tag{2.1a}$$

If $\varphi \in C_0^\infty(a,b)$ then $\varphi(a) = \varphi(b) = 0$ and $[u\varphi]_a^b = 0$, then

$$\int_a^b u'\varphi dx = -\int_a^b u\varphi' \, dx \text{ holds for all } u \in C^1(a,b), \, \varphi \in C_0^\infty(a,b) \tag{2.1b}$$

The generalization of these two formulas to higher dimensions use the classical Green's theorem provided Ω is a nonempty bounded set in $\mathbb{R}^n$ that has sufficiently smooth boundary

$$\int_\Omega \frac{\partial u}{\partial x_j}\varphi d\mathbf{x} = \int_{\partial\Omega} u\varphi n_j dS - \int_\Omega u\frac{\partial\varphi}{\partial x_j}d\mathbf{x} \text{ For all } u,\varphi \in C^1(\bar{\Omega}), \tag{2.2a}$$

where n_j is the j-th component of the outward unit normal vector n to the boundary $\partial\Omega$ of the domain Ω. Replacing in (2.2a), by a function φ in $C_0^\infty(\mathbb{R}^n)$ and using the fact $\varphi = 0$ on $\partial\Omega$, the result is

$$\int_\Omega u\frac{\partial\varphi}{\partial x_k}dx = -\int_\Omega \varphi\frac{\partial u}{\partial x_k}dx, \text{ holds for all } u \in C^1(\Omega) \text{ and } \varphi \in C_0^\infty(\Omega). \tag{2.2b}$$

Definition 10 To define distributions we provide to $\mathcal{D}(\mathbb{R}^n)$, the following notion of convergence: If $\{\varphi_k\}$ is a sequence in $\mathcal{D}(\mathbb{R}^n)$, is said that $\varphi_k \to \varphi$ in $\mathcal{D}(\mathbb{R}^n)$ if (i) $\partial^\alpha \varphi_k \to \partial^\alpha \varphi$ uniformly for all multi-indices α and (ii) the φ_k's and φ are all supported in a common compact set (for more details see [8, 9]).

A **distribution** on $\mathbb{R}^n$ is a linear functional $T : C_0^\infty(\mathbb{R}^n) \to \mathbb{R}$, that is continuous in the sense that if $\varphi_k \to \varphi$ in $C_0^\infty(\mathbb{R}^n)$ then $T[\varphi_k] \to T[\varphi]$. The space of distributions is denoted by $\mathcal{D}'(\mathbb{R}^n)$ or $\mathcal{D}'$.

A distribution (or generalized function) is characterized by its "actions" $T[\varphi]$ or $\langle T, \varphi \rangle$ ("Duality bracket") over the elements $\varphi \in \mathcal{D}(\mathbb{R}^n)$. For instance $\langle \delta, \varphi \rangle = \varphi(0)$ is the action of the Dirac delta functional over φ. In general each $a \in \mathbb{R}^n$ determines a linear functional $\delta_{(a)} = \delta(x-a)$ on $\mathcal{D}(\mathbb{R}^n)$ by the expression $\langle \delta_{(a)}, \varphi \rangle = \varphi(a)$. This is a way to solve the formal inconsistency settled by Dirac's sampling property.

An important thing to note is that any locally integrable function f, will define a distribution by

$$\langle f, \varphi \rangle = \int_{\mathbb{R}^n} f(x)\varphi(x)dx \; \forall \varphi \in C_0^\infty(\mathbb{R}^n) \tag{2.1}$$

These are called **regular** distributions. If this is not the case, they are called **singular** distributions (for example, δ). By abuse of notation singular distributions are also denoted by the symbol $f(x)$ used for ordinary functions, although there is no value at the point x and the effect of an arbitrary distribution f on φ is written as an integral $\int f(x)\varphi(x)dx$. So we commonly write $\int \delta(x)\varphi(x)dx$ as equivalent to $\langle \delta, \varphi \rangle$, and we proceed as in ordinary calculus, for example

$$\int_{\mathbb{R}} \delta(x-a)\varphi(x)dx = \int \delta(\xi)\varphi(\xi+a)d\xi = [\varphi(\xi+a)]_{\xi=0} = \varphi(a)$$

Operations on functions can be extended to distributions (derivatives, convolution, Fourier transform). One key idea is that the definition of operations on distributions should coincide with the definition for regular distributions. For example, from $\int (f\xi)\varphi = \int f(\xi\varphi)$, it follows that the product of a distribution T in $\mathcal{D}'(\mathbb{R}^n)$ and a function $\xi \in C_0^\infty(\mathbb{R}^n)$ is defined as

$$\langle \xi T, \varphi \rangle = \langle T, \xi\varphi \rangle.$$

Nevertheless, this should be done carefully, because it may arise several limitations; for example, it will not be possible to multiply distributions nor define the Fourier transform without making extra assumptions. It is a well-known fact that classical function spaces may not be suitable in order to formulate well-posedness. The derivative concept is in the core of this difficulty, so it is necessary to get a more versatile definition of derivative.

Distributional derivatives are motivated by integration by parts formulas (2.1b)

$$\int_a^b u'\varphi dx = -\int_a^b u\varphi' \, dx \;\; \text{holds for all } u \in C^1(a,b), \; \varphi \in C_0^\infty(a,b)$$

Thus, in one variable we may consider the functional $\langle f', \varphi \rangle = \int_{-\infty}^\infty f'\varphi$ and applying integration by parts

$$\langle f', \varphi \rangle = \int_{-\infty}^\infty f'\varphi = -\int_{-\infty}^\infty f\varphi'$$
$$= -\langle f, \varphi' \rangle$$

Then the derivative of the functional $\langle f, \varphi \rangle$ is the functional $-\langle f, \varphi' \rangle$. Thus, given any distribution f, it is defined

$$\left\langle \frac{\partial f}{\partial x_k}, \varphi \right\rangle = -\left\langle f, \frac{\partial \varphi}{\partial x_k} \right\rangle$$

and applying this repeatedly

$$\int_\Omega (\partial^\alpha f)\varphi dx = (-1)^{|\alpha|} \int_\Omega f(\partial^\alpha \varphi)dx.$$

Assuming sufficient conditions of differentiability of u, we can define a regular distribution

$$\langle \partial^\alpha f, \varphi \rangle = (-1)^{|\alpha|}\langle f, \partial^\alpha \varphi \rangle \quad \forall \varphi \in C_0^\infty(\mathbb{R}^n) \tag{2.3}$$

This means that the derivative of a non-differentiable function can be defined in terms of relations with smooth functions of compact support. For example, the derivative δ' of the delta distribution it is expressed as $\langle \delta', \varphi \rangle = -\langle \delta, \varphi' \rangle = \varphi'(0)$. By (2.3), a distribution T has derivatives of all orders, a direct consequence is that T is indefinitely differentiable and always

$$\frac{\partial^2 T}{\partial x_j x_k} = \frac{\partial^2 T}{\partial x_k x_j}$$

This point of view solves the famous problem about quantities such as Dirac delta and Heaviside step function do not have derivatives in the classical sense. Nevertheless, treating them as distributions allows to extend the concept of a derivative in such a way that any number of derivatives can be defined for these quantities and, even, for any distribution. Thus, the classical notion of a derivative is recovered.

Example 4 The derivative of the Heaviside function $H(x) = 0$ if $x < 0$, $H(x) = 1$ if $x > 0$ is $H' = \delta$. This can be seen as

$$\langle H', \varphi \rangle = -\langle H, \varphi' \rangle = -\int_0^\infty \varphi'(x)dx = \varphi(0) - \varphi(\infty) = \varphi(0) = \langle \delta, \varphi \rangle.$$

Example 5 The derivative of a function f with jumps can be calculated as a distribution. Suppose f is a piecewise function on $\mathbb{R}$ and differentiable at all $x \neq 0$ but has a jump discontinuity at $x = 0$. Then, calling f^1 the pointwise derivative and f' the distributional derivative, we have

$$\langle f', \varphi \rangle = -\int f(x)\varphi'(x)dx = \int_{-\infty}^0 f\varphi'dx - \int_0^\infty f\varphi'dx$$

$$= -f(x)\varphi(x) \Big|_{-\infty}^0 + \int_{-\infty}^0 f^{[1]}\varphi dx - f(x)\varphi(x) \Big|_0^\infty + \int_0^\infty f^{[1]}\varphi dx$$

$$= f(0-)\varphi(0) + f(0+)\varphi(0) + \int_{-\infty}^\infty f^{[1]}\varphi dx,$$

thus,

$$f' = f^1 + [f(0+) - f(0-)]\delta$$

2.4.1 Convolution

Many important results of applied mathematics can be expressed as convolutions. Given two functions f and u their **convolution product** is a new function $f * u(x)$ such that

$$f * u(x) = \int_{\mathbb{R}^n} f(x-t)u(t)dt$$

This definition can be generalized to distributions [8, 9, 59, 60] and several conditions on f and u are necessary to ensure that the integral exists. Convolution obeys the same algebraic laws of ordinary multiplication (i) $f * (\alpha u + \beta v) = \alpha(f * u) + \beta(f * v)$, (ii) $f * u = u * f$, (iii) $f * (u * v) = (f * u) * v$. One property which convolution does not share with ordinary multiplication is that, on the contrary to $f * 1 = f$ for all f, there is no function u such that $f * u = f$. This limitation is solved introducing distribution theory, where $f * \delta = f$; because

$$\delta * f(x) = \int \delta(t)f(x-t)dt = f(x-t)\big|_{t=0} = f(x)$$

Very useful for function approximation is the existence of sequences $\{u_k\}$ such that $f * u_k$ converges to f as $n \to \infty$. In these problems convolution behaves as a continuous superposition of translates of f and $f * u_k$ may be regarded as a smoothed version of f. In the next chapters we will see how splines approximators are expressed as some kind of convolution with fundamental solutions of differential operators.

The convolution of a distribution $T \in D'(\mathbb{R}^n)$ with a test function $\varphi \in \mathcal{D}(\mathbb{R}^n)$ is a C^∞ function $T * \varphi$ on $\mathbb{R}^n$. $T * \varphi$ is called the **regularization** of T. Besides the above properties, convolution has the following properties with Dirac's delta

(i) $\delta * T = T$

(ii) $\delta_{(a)} * T = \tau_a T$

(iii) $\delta' * T = T'$

(iv) $\delta_{(a)} * f(x) = f(x-a)$ (2.4)

In general, if D is a differential operator with constant coefficients in $\mathbb{R}^n$, $D\delta * T = DT$. For example, if D is the Laplacian operator $\Delta = \sum_{i=1}^{n} \dfrac{\partial^2}{\partial x_i^2}$ in $\mathbb{R}^n$, then $\Delta\delta * T = \Delta T$.

2.4.2 The Schwartz Space and Fundamental Solutions

A very important problem when extending the Fourier transform of a function f

$$\mathcal{F}[f] = \hat{f}(\xi) = \int_{\mathbb{R}^n} f(x)e^{-ix\cdot\xi}dx, \ \xi = (\xi_1, \xi_2, \ldots, \xi_n),$$

to distributions is that if $\varphi \in \mathcal{D}$ it is possible that $\hat{\varphi} \notin \mathcal{D}$. In order to obtain a useful definition of Fourier transform, Schwartz defined the space $\mathcal{S}$ of functions $\varphi \in C^{\infty}$ such that φ and all its derivatives $\partial^{\alpha}\varphi$ vanish at infinity more rapidly that any power of $\|x\|$. For example $p(x)e^{-\|x\|^2}$ belongs to $\mathcal{S}$ with p any polynomial. With this in mind, the space of tempered distributions is defined as the dual space $\mathcal{S}'(\mathbb{R}^n)$. As a consequence $\varphi \in \mathcal{S}$ implies $\hat{\varphi} \in \mathcal{S}$ and it is possible to define $\langle \hat{T}, \varphi \rangle = \langle T, \hat{\varphi} \rangle$, $\forall T \in \mathcal{S}'$; preserving in this way, the well-known nice properties of Fourier transform ($\hat{\delta} = 1$, for example).

A *fundamental solution* of a linear differential operator

$$P = \sum_{|\alpha| \leq m} c_{\alpha} \partial^{\alpha}$$

with constant coefficients c_{α}, is a distribution $K(\mathbf{x}, \mathbf{y})$, such that

$$P_{\mathbf{x}}(K(\mathbf{x}, \mathbf{y})) = \delta(\mathbf{x} - \mathbf{y}).$$

An important example is the fundamental solutions of the Laplacian operator which are the solutions of the differential equation $\Delta K = \delta$. In this case K is called a potential. Fundamental solutions have the remarkable property of being part of the solution of inhomogeneous differential equations of the form $Pu = f$. This is shown in the following way: if $PK = \delta$ then

$$P(K * f) = (PK) * f = \delta * f = f,$$

therefore

$$K * f \text{ is a solution of } Pu = f. \tag{2.5}$$

The next chapters show the significant value of this property for multivariate approximation. The interpolants we are going to construct are expressed as a linear combination of translates of the fundamental solution for a differential operator. Fundamental solutions are called Green's functions when they are subjected to boundary conditions. According to Malgrange-Ehrenpreis theorem [9], every operator $P = \sum_{|\alpha| \leq m} c_{\alpha} \partial^{\alpha}$, has a fundamental solution. Fundamental solutions take different names depending of the specific field of application: impulse response, Green's functions or influence functions.

Example 6. Fundamental solutions of iterated Laplacian Δ^m [20]. If K is a fundamental solution of the operator Δ^m, then $\Delta^m K = \delta$. Taking Fourier transform on both sides, $(-1)^m \rho^{2m} E = 1$. Using formulas (2.4), are obtained the fundamental solutions in the form of radial function, i.e. depends of the Euclidian distance $r = \|x - y\|$.

$$K(x,y) = \Phi_{m,n}(\| x - y \|) \qquad \Phi(r) = \begin{cases} cr^{2m-n} \ln r, n \text{ even} \\ dr^{2m-n}, n \text{ odd} \end{cases}$$

2.5 Sobolev Spaces

The well-known classical spaces, that is the spaces of continuous functions, may fail when dealing with ill-posed problems. Distribution theory gives a setting to construct spaces where these problems becomes well posed. The space of distributions $\mathcal{D}'(\mathbb{R}^n)$ as described above provides answers for conditions of existence and uniqueness. However, this space is "very large", therefore, the issues on regularity are treated into some of its subspaces. It is possible to reconstruct the space $\mathcal{L}^2(\mathbb{R}^n)$ as a Hilbert space of distributions, unifying this theory with the theory of $\mathcal{L}^2(\mathbb{R}^n)$ spaces; this leads naturally to Sobolev spaces, very convenient for the pure and numerical treatment.

The key idea of Sobolev techniques it is to assume that the distributions which solve a particular problem really come from a function f, but without making any smoothness assumption about f. The next step in the method, is to take advantage of the operational properties of distributions to solve the problem. Thus it is possible to use the so called Sobolev embedding properties in order to determine the smoothness degree of the solution. If m is a nonnegative integer and $p \in [1, \infty]$, The Sobolev space $W^{m,p}(\mathbb{R}^n)$, is the vector space

$$W^{m,p}(\mathbb{R}^n) = \{u \in \mathcal{D}' : \partial^\alpha u \in \mathcal{L}^p(\mathbb{R}^n), |\alpha| \le m\},$$

consisting of those functions in $\mathcal{L}^p(\mathbb{R}^n)$ that, together with all their distributional partial derivatives up to and including those of order m, belong to $\mathcal{L}^p(\mathbb{R}^n)$. This space is provided with the norm

$$\| u \|_{m,p} = \left(\sum_{|\alpha| \le m} \int_{\mathbb{R}^n} | \partial^\alpha u |^p \, dx \right)^{1/p}$$

The spaces $W^{m,2}(\mathbb{R}^n)$ ($p = 2$) are symbolized as $H^m(\mathbb{R}^n)$ and consists of those functions in $\mathcal{L}^2(\mathbb{R}^n)$ that, together with all their distributional derivatives of order $|\alpha| \le m$, belong to $\mathcal{L}^2(\mathbb{R}^n)$

$$H^m(\mathbb{R}^n) = \{u \in \mathcal{D}' : \partial^\alpha u \in \mathcal{L}^2(\mathbb{R}^n), |\alpha| \le m\}$$

We consider real value functions only, and make $H^m(\mathbb{R}^n)$ an inner product space with the Sobolev Inner product

$$(u,v) = \sum_{|\alpha| \le m} \int_{\mathbb{R}^n} (\partial^\alpha u)(\partial^\alpha v) dx; u, v \in H^m,$$

that generates the Sobolev norm $\| u \|_{H^m}^2 = \int_{\mathbb{R}^n} \sum_{|\alpha| \le m} (\partial^\alpha u)^2 dx$.

Thus, $H^m(\mathbb{R}^n)$ is a Hilbert space with respect to this norm. With the above setting, distribution theory provides a very efficient tool for dealing with complex problems; nevertheless, it is very important to remember that in order to have a realistic application it should be done with spaces of continuous functions. This is the method used in partial differential equations, where the problems are first solved in the realm of distributions and then is verified if these distribution solutions are classical solutions. Sobolev embedding theorems answers this question. By the *Sobolev embedding theorem* [9], it is well known that for $m > \dfrac{n}{2}$, the inclusion $H^m(\mathbb{R}^n) \subseteq C(\mathbb{R}^n)$ holds, or, to be more precise, that every equivalence class in $H^m(\mathbb{R}^n)$ contains a continuous representer. In this way, $H^m(\mathbb{R}^n)$ is interpreted as a set of continuous functions. The next chapter shows some practical applications of the abstract framework stated here.

Chapter 3
Variational Methods

Variational calculus study the optimization of functionals, inspired on the differential calculus of functions. Under this viewpoint can be obtained the well-known Euler-Lagrange conditions for extremals of functionals. This can also be deduced by the Gateaux variation, a generalization of the directional derivative of a function. Once we have developed this valuable tool it is possible to solve some problems of Regularization theory.

3.1 Introduction

Regularization methods are based in variational principles that have been applied with considerable success in computer vision and other fields of modern technology where it is necessary the recovery of an object from a set of data. This is a task common to many areas of imaging such as pattern recognition and medical diagnosis.

During the second half of the twentieth century many branches of applied mathematics aroused motivated by the problems of computer vision, resulting in tight relationships between approximation theory, splines and functional analysis. Different arguments suggest that early vision processes correspond to conceptually independent modules classified as edge detection, surface reconstruction, spatio-temporal interpolation, among others. The mathematical theory developed for regularizing ill-posed problems leads in a natural way to the solution of early vision problems in terms of variational principles.

One of the first references to applications of variational theory in interpolation is a Schoenberg paper of 1964 [57, 58]. This paper described the problem as follows: Given the knots $-\infty \leq a < x_1 < x_2 < \cdots < b \leq \infty$, the problem is to find a function f in the Sobolev space W^m of functions with $m-1$ continuous derivatives and m-th derivative square integrable, to minimize $\int_a^b (f^{(m)}(x))^2 dx$, such that $f(x_i) = y_i$, $i = 1, 2, \ldots, n$. The solution is the well-known cubic polynomial spline that will be studied in the next chapter. This point of view was then generalized to several variables; for example, Duchon [11–14] applied a general approach based on distribution spaces

H. Montegranario, J. Espinosa, *Variational Regularization of 3D Data*,
SpringerBriefs in Computer Science, DOI 10.1007/978-1-4939-0533-1_3,
© The Author(s) 2014

and obtained multidimensional splines which are a natural generalization to $\mathbb{R}^n$ of Schoenberg spline (thin plate spline), in terms of radial basis functions. By the Sobolev embedding theorem, he found that his spaces are included into the space of continuous functions, making possible the work of interpolation. The method introduced by Duchon followed the ideas of Attéia [3] and Laurent [36], for the general theory of Hilbertian splines and involves reproducing kernels theory. For more details on the historical development see [16].

Currently there exist a great number of applications that has been made possible by the advent of computers with enough computing power to deal with the large amounts of image data and the complexity of the algorithms that operate on them. However we will see in the coming sections that variational calculus was created long before the technical and computer revolution era.

3.2   Extremals of Functionals

Variational calculus was born in the spirit of natural philosophy of the European enlightenment of 18th century, so it is very useful (and impossible not to do it) to introduce the subject mentioning some aspects of its historical background. From its initial conception to the present day, variational calculus has been a source of inspiration and applications for all kind of scholars. For example, current philosophers and humanists would be interested in some metaphysical ideas of ancient Greeks and eighteenth century encyclopedists that gave inspiration to some initial problems. Scientists will find that Hamilton's principle of least action can be made to develop the basic laws of electricity and magnetism, quantum theory and relativity and modern computer scientist would see in variational calculus the possibility to solve optimization problems in so diverse applications such as medical imaging, inverse theory, digital elevation models for geosciences data and many others.

Variational calculus developed from the ideas about determination of extrema of functions in the classical infinitesimal calculus of Newton and Leibnitz. These ideas were first applied to functionals involving an unknown function and boundary conditions, in the form

$$J[y] = \int_{x_0}^{x_1} F(x, y, y')dx, \ y(x_0) = y_0, \ y(x_1) = y_1, \ x_0 < x_1 \tag{3.1}$$

The most famous of these problems (proposed by Johann Bernoulli in 1696) had to do with minimizing the brachystochrone curve, defined by the functional

$$J[y] = \int_{x_0}^{x_1} \frac{\sqrt{1 + y'^2}}{\sqrt{2gy}} dx.$$

The minimum of this functional, determines the curve along which a particle will fall from one given point to another in the shortest time, a problem solved by Newton, Leibniz, and Johann Bernoulli as well as by his brother Jacob (1654–1705), the solution being a cycloid.

Given a functional $J[y]$ defined in a subset $\mathcal{G}$ of a function space $\mathcal{U}$ the goal of variational calculus is to find the elements $y_0 \in \mathcal{G}$ such that either

$$J[y] \geq J[y_0] \text{ for all } y \in \mathcal{G}; (\ y_0 \text{ is a minimum})$$

or

$$J[y] \leq J[y_0] \text{ for all } y \in \mathcal{G}; (y_0 \text{ is a maximum})$$

Since the condition for maximum is equivalent to $-J[y] \geq -J[y_0]$, then it is sufficient to develop the theory for minimums. If there exist only one function y_0, that minimizes $J[y] \forall y \in \mathcal{G}$, then y_0 is called a global minimum. Given that in these problems we do not have a function of finitely many independent real variables but a functional $J[y]$ on a class of functions, this process is called **infinite-dimensional optimization**. Other disciplines which study infinite-dimensional optimization are optimal control and shape optimization [40].

3.3 Euler-Lagrange Equation

The first general result about extremals of functionals was given by Lagrange who discovered the well-known necessary condition that bears his name. The demonstration of this theorem resumes in some few lines the classical techniques of variational calculus and predicts its modern development; we will give some of its details. First we quote one version of the so called fundamental lemma of variational calculus:

Lemma 1 If $f(x)$ is continuous in $[a,b]$, and if $\int_a^b f(x)\varphi(x)dx = 0$ for every $\varphi(x)$ in the set $D_0 = \{\varphi \in C^1[a,b]: \varphi(a) = \varphi(b) = 0\}$, then $f(x) = 0$ for every $x \in [a,b]$.

In order to find the minimum of the functional $J[y] = \int_{x_1}^{x_2} F(x, y, y')\, dx$, we assume $y(x)$ as a minimizer and that everything is sufficiently differentiable for our purposes.

Now, consider the set of functions $\varphi \in D_0$ such that $\varphi(x_1) = \varphi(x_2) = 0$; if ε is a small parameter, then $\bar{y}(x) = y(x) + \varepsilon\varphi(x)$ represents a one parameter family of admissible functions that satisfy the boundary conditions, because $\bar{y}(x_i) = y(x_i)$, $i = 1, 2$. We consider $J[y + \varepsilon\varphi]$ as a function Φ in the variable ε, then

$$\Phi(\varepsilon) = \int_{x_1}^{x_2} F[x, y(x) + \varepsilon\varphi(x), y'(x) + \varepsilon\varphi'(x))]dx.$$

This expression reduces the problem to a question in one-variable calculus; so the minimum is found by making

$$\frac{\partial \Phi}{\partial \varepsilon}\bigg|_{\varepsilon=0} = 0,$$

then applying the chain rule

$$\frac{\partial \Phi}{\partial \varepsilon} = \int_{x_1}^{x_2} \frac{\partial}{\partial \varepsilon} F(x, \overline{y}, \overline{y}') dx = \int_{x_1}^{x_2} \frac{\partial F}{\partial \overline{y}} \varphi + \frac{\partial F}{\partial \overline{y}'} \varphi'$$

and taking $\varepsilon = 0$

$$\frac{\partial \Phi}{\partial \varepsilon}\bigg|_{\varepsilon=0} = \int_{x_1}^{x_2} F_y \varphi + F_{y'} \varphi' = 0$$

Integrating by parts the second term in the former integral

$$\int_{x_1}^{x_2} F_{y'} \varphi' = [F_{y'} \varphi]_{x_1}^{x_2} - \int_{x_1}^{x_2} \frac{d}{dx} F_{y'} \varphi = 0 - \int_{x_1}^{x_2} \frac{d}{dx} F_{y'} \varphi,$$

then

$$\int_{x_1}^{x_2} F_y \varphi - \frac{d}{dx} F_{y'} \varphi = \int_{x_1}^{x_2} \left[F_y - \frac{d}{dx} F_{y'} \right] \varphi = 0, \text{ for all } \varphi \in D_0,$$

and by the fundamental lemma, the final result is the necessary condition well-known as **Euler-Lagrange equation**.

$$\frac{\partial F}{\partial y} - \frac{d}{dx} \left(\frac{\partial F}{\partial y'} \right) = 0, \tag{3.2}$$

The variational lemma used in this proof was considered self-evident for Lagrange and his contemporaries, until a rigorous demonstration was given by Du Bois-Reymond (1879 [40]). The idea of using integration by parts and the functions $\varphi(x)$ that vanish in the ends of the interval $[x_1, x_2]$, eventually led to the creation of the theory of distributions. The expression

$$\frac{\delta J}{\delta y(x)} \equiv \frac{\partial F}{\partial y} - \frac{d}{dx} \left(\frac{\partial F}{\partial y'} \right),$$

is sometimes called **functional derivative** or **Frechet derivative** of $J[y]$ with respect to y.

Example 1 (Minimal surfaces of revolution). One of the questions that has generated great interest from Euler times up to now, is the problems of minimal surfaces, which has found many applications in surface design and computer aided manufacturing (CAM).

Fig. 1.1 The catenoid. A
surface of revolution with
minimal area

A simple version of the problem is to find the function $y(x)$ going through the points $P_0(x_0, y_0)$ and $P_1(x_1, y_1)$ that generates the surface of revolution with minimal surface. The corresponding variational problem is

$$J[y] = 2\pi \int_{x_0}^{x_1} y\sqrt{1 + y'^2}\, dx = \min,$$ with boundary conditions $y(x_0) = y_0$, $y(x_1) = y_1$

Applying Euler equation $y\sqrt{1 + y'^2} - \dfrac{yy'^2}{\sqrt{1 + y'^2}} = c_0$, from which we obtain

$$x = c_0 \int \frac{1}{\sqrt{y^2 - c_0^2}}\, dy$$

$$x = c_0 \cosh^{-1}\left(\frac{y}{c_0}\right) + c_1$$

And finally the solution is the catenoid

$$y = c_0 \cosh^{-1}\left(\frac{x - c_1}{c_0}\right)$$

Example2 Using the parametric version of the catenoid in matlab, we can easily obtain a graph of surface (Fig. 1.1).

```
>> ezsurf('4*cos(u).*cosh(v/4)', '4*sin(u).*cosh(v/4)', 'v')
```

Variational problems do not always have a solution, Euler–Lagrange formula is only a necessary condition for minimum, the solution y could also produce a maximum or an inflexion point for $J[y + \varepsilon\varphi]$ at $\varepsilon = 0$. To define the nature of the extremum is necessary to study the sign of higher derivatives of $J[y]$. Although this can be done, it could become complicated. Fortunately, for most applications, Euler–Lagrange equation by itself is enough to give a complete solution of the problem, because the existence and the nature of an extremum are often clear from the physical or geometrical meaning of the problem. Wierstrass [40] was the first to point out that in some cases, a minimizing function can come arbitrarily close to the lower bound without ever reaching it. The methods that use Euler-Lagrange equation to reduce variational problems to differential equations are called **indirect methods** because they are based in necessary conditions, without establishing its existence. On the other side, there exist the **direct methods**, that proceed directly to the minimization of the given functional without first looking at necessary conditions.

3.4   Generalization of Euler-Lagrange Equation

Lemma 3 (The variational lemma in $C_0^\infty(\Omega)$ [73]). Let $\Omega \subset \mathbb{R}^n$, then it follows from $u \in \mathcal{L}^2(\Omega)$ and

$$\int_\Omega u\varphi dx = 0 \text{ for all } \varphi \in C_0^\infty(\Omega),$$

that $u(x) = 0$ for almost all $x \in \Omega$. If in addition $u \in C(\Omega)$, then $u(x) = 0$ for all $x \in \Omega$.

In general, a variational problem may be of the form

$$\int_\Omega F(x,u,Du)\,dx = \min! \, u \in C^{2m}(\overline{\Omega})$$

$$D^\alpha u = g_\alpha \text{ for all } |\alpha| \le m-1 \tag{3.3}$$

In this connection we set $u = (u_1,...,u_k)$. The Lagrange function F depends on u and all partial derivatives of u up to order m; furthermore, $D^\alpha u$ denotes an arbitrary partial derivative of order $|\alpha|$. Let $Du = (D^\alpha u)_{1 \le |\alpha| \le m}$

Theorem 1 (Euler-Lagrange general form). Let Ω be a bounded region in $\mathbb{R}^n$ with $n \ge 1$ and suppose that the Lagrange function $F : \Omega \times \mathbb{R}^{K+M} \to \mathbb{R}$ is C^2. Then each solution u of the variational problem (3.3) is also a solution of the Euler-Lagrange equations

$$\sum_{|\alpha| \le m} (-1)^{|\alpha|} D^\alpha \left[\frac{\partial F}{\partial D^\alpha u_k}(\mathbf{x}, u(x), Du(\mathbf{x})) \right] = 0 \quad k = 1,2,...,K \tag{3.4}$$

Moreover, u is also a solution of the generalized Euler-Lagrange equations

$$\sum_{|\alpha| \le m} \int_\Omega \frac{\partial F}{\partial D^\alpha u_k}(\mathbf{x}, u(x), Du(\mathbf{x})) D^\alpha \varphi_k(\mathbf{x})\, d\mathbf{x} = 0 \text{ For all } \varphi_k \in C_0^\infty(\Omega) \text{ and all } k = 1,...,K$$

Example 3 The variational properties of the thin plate functional

$$J[u(x, y)] = \int_\Omega (u_{xx}^2 + 2u_{xy}^2 + u_{yy}^2) \, dy \, dx,$$

have been widely studied in computer vision problems, especially surface reconstruction. By (3.4) this functional has the general form

$$J[u] = \int_\Omega F(x, u, u_x, u_y, u_{xx}, u_{xy}, u_{yy}) \, dy \, dx$$

with the functional derivative

$$\frac{\delta J}{\delta u} = F_u - \frac{\partial}{\partial x} F_{u_x} - \frac{\partial}{\partial y} F_{u_y} + \frac{\partial^2}{\partial x^2} F_{u_{xx}} + \frac{\partial^2}{\partial x \partial y} F_{u_{xy}} + \frac{\partial^2}{\partial y^2} F_{u_{yy}}$$

$$= \frac{\partial}{\partial x^2} (2u_{xx}) + \frac{\partial}{\partial x \partial y} (4u_{xy}) + \frac{\partial}{\partial y^2} (2u_{yy})$$

$$= 2\left(\frac{\partial^4 u}{\partial x^4} + 2\frac{\partial^4 u}{\partial x^2 \partial y^2} + \frac{\partial^4 u}{\partial y^2} \right)$$

$$= 2\Delta(\Delta(u)),$$

where Δ is the Laplacian operator $\Delta u = u_{xx} + u_{yy}$. The operator $\Delta(\Delta u) = \Delta^2 u$, called *biharmonic*, has great importance in elasticity theory

3.5 The Gateaux Variation

The Gateaux differential is defined for a transformation (possibly nonlinear) $J : \mathcal{G} \subset \mathcal{U} \to \mathcal{V}$, with $\mathcal{U}$ a vector space and $\mathcal{V}$ a normed space. The space $\mathcal{U}$ is not necessarily normed. This differential generalizes the familiar concept of directional derivative in n-dimensional Euclidian space.

Definition Let J be the transformation $J : \mathcal{G} \subset \mathcal{U} \to \mathcal{V}$, $y \in \mathcal{G}$ and let φ be arbitrary in $\mathcal{U}$. Then the limit

$$\delta(y; \varphi) := \lim_{\varepsilon \to 0} \frac{J[y + \varepsilon\varphi] - J[y]}{\varepsilon},$$

if exists, it is called the Gateau Differential of J at y in the direction of φ. If this limit exists for each $\varphi \in \mathcal{U}$, the transformation is called Gateau differentiable at y and $\delta J(y, \varphi)$ is called the first variation of the functional $J[\cdot]$. The most important case of this definition is $\mathcal{V} = \mathbb{R}$, $J[\cdot]$ is then a real-valued functional. The Gateaux variation of $J[\cdot]$ at y depends only on the local behavior of $J[\cdot]$ near y; but this variation may not exist in any direction φ different from zero or it may exist in some directions and not in others. The Euler condition can be derived from this definition and the following theorem.

Theorem 2 Let $J[\cdot]$ be a real valued functional with Gateaux differential $\delta J(y, \varphi)$ on the vector space $\mathcal{U}$. A necessary condition for $J[\cdot]$ to have an extremum at $y_0 \in \mathcal{U}$ is that $\delta J(y, \varphi) = 0$ for all $\varphi \in \mathcal{U}$.

This result follows from observing that $J[y + \varepsilon\varphi]$ is a function of the real variable ε with extremum in $\varepsilon = 0$. Then by single variable calculus

$$\delta J(y, \varphi) = \frac{\partial}{\partial \varepsilon} J[y + \varepsilon\varphi]\big|_{\varepsilon=0} = 0 \tag{3.5}$$

Example 4 The minimum distance between two points $P_0(x_0, y_0)$ and $P_1(x_1, y_1)$ in the plane is a line segment. This can be seen minimizing the arc-length functional

$$J[y] = \int_{x_0}^{x_1} \sqrt{1 + y'^2}\, dx$$

calculating the Gateaux differential $\delta J(y, \varphi) = 0$,

$$J[y + \varepsilon\varphi] = \int_{x_0}^{x_1} \sqrt{1 + (y' + \varepsilon\varphi')^2}$$

$$\frac{\partial [Jy + \varepsilon\varphi]}{\partial \varepsilon} = \int_{x_0}^{x_1} \frac{\partial}{\partial \varepsilon} \sqrt{1 + (y' + \varepsilon\varphi')^2}\, dx$$

$$\delta J(y, \varphi) = \int_{x_0}^{x_1} \frac{y'}{\sqrt{1 + y'^2}} \varphi'\, dx$$

assuming $u = \dfrac{y'}{\sqrt{1 + y'^2}}$ and applying integration by parts

$$\delta J(y, \varphi) = \int u\varphi' = -\int u'\varphi = 0 \; \forall \varphi \in C_0^\infty(x_0, x_1),$$

Now, by the fundamental lemma $u' = 0$, thus $\dfrac{d}{dx} \dfrac{y'}{\sqrt{1 + y'^2}} = 0$ and as a consequence $\dfrac{y'}{\sqrt{1 + y'^2}} = const.$

Simplifying this last expression, $y' = c$, and integrating we obtain the result we were expecting, the equation of a line $y = cx + d$.

3.6   Surfaces and Variational Properties

A surface Σ can be seen as a collection of points in $\mathbb{R}^3$ with several forms of representation:

- *Graph representation.* As a set of points with the form $(x, y, f(x, y))$, where $z = f(x, y)$ is a smooth function of two variables.
- *Implicit representation.* As the set of all points (x, y, z) that satisfy an equation of the form $g(x, y, z) = 0$; where g is also a smooth function
- *Parametric representation.* With two parameters
- $\mathbf{x}(u_1, u_2) = (x_1(u_1, u_2), x_2(u_1, u_2), x_3(u_1, u_2))$

where u_1, u_2 range over some domain Ω in the plane $\mathbb{R}^2$ and $\mathbf{x} : \Omega \to \mathbb{R}^3$ is a smooth map. Note that a parametric representation $\mathbf{x}(u_1, u_2)$ is really a mapping but it is used to identify the mapping with its image Σ. Vectors $\mathbf{x}_1 = \dfrac{\partial \mathbf{x}}{\partial u_1}$ and $\mathbf{x}_2 = \dfrac{\partial \mathbf{x}}{\partial u_2}$ are tangent at every point p of Σ. On this point there exist a tangent plane $T_p \Sigma$ generated by the vectors $\mathbf{x}_1, \mathbf{x}_2$, it is to say $T_p \Sigma = span\{\mathbf{x}_1, \mathbf{x}_2\}$. Taking the vector product as $\mathbf{x}_1 \wedge \mathbf{x}_2$, is obtained the unit normal vector function (Gauss map) $\hat{\mathbf{n}} : \Omega \to S^2$ where

$$\breve{\mathbf{n}} = \frac{\mathbf{x}_1 \wedge \mathbf{x}_2}{\|\mathbf{x}_1 \wedge \mathbf{x}_2\|}$$

and S^2 is the unitary sphere. It is possible to demonstrate that at each point p of a surface Σ there exist two particular directions of the tangent pane $T_p \Sigma$ such that:

- They are mutually perpendicular
- The curvatures κ_1 and κ_2 of the normal sections in these directions are the smallest and largest values of the curvatures of all normal sections
- The curvature $\kappa(\theta)$ of the normal section rotated from the section with curvature κ_1 by the angle θ is expressed by the formula

$$\kappa(\theta) = \kappa_1 \cos^2 \theta + \kappa_2 \sin^2 \theta,$$

such directions are called the principal directions and the curvatures κ_1 and κ_2 are called the **principal curvatures** of the surface at the given point.

The shape or complexity of a surface can be described by its curvature. Many problems in imaging can be studied using functionals based on the concept of curvature. Two classical methods to measure curvature are defined using principal curvatures κ_1, κ_2. The first is

$$\text{Gaussian curvature } K = \kappa_1 \kappa_2,$$

the other is

$$\text{Mean curvature } H = \frac{1}{2}(\kappa_1 + \kappa_2).$$

These couple of simple expressions have relevant consequences for the analysis of surfaces. The sign of the Gaussian curvature defines the character of the surface near the point under consideration. For $K > 0$ the surface has the form of a bowl

Fig. 1.2 An example of
minimal surface

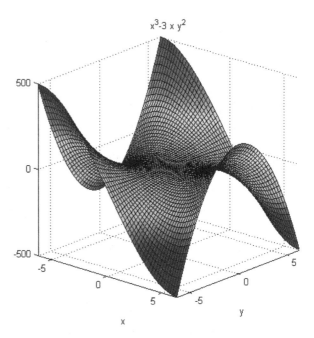

(κ_1 and κ_2 have the same sign) and for K < 0, when κ_1 and κ_2 have different sign,
the surface is like a saddle (Fig. 1.2).

A way to obtain very useful formulas for curvatures is the following. Consider the tangent plane $T_p\Sigma$ to the surface at the point $p \in \Sigma$ defined as $z = f(x, y)$. For convenience of the analysis [51], we can obtain a proper orthogonal change of coordinates and a translation under which the z-axis is perpendicular to $T_p\Sigma$ and $p = (0,0,0)$ then f_x and f_y vanish at p and by Taylor series

$$f(h,k) = f(p) + \frac{1}{2}\left(h^2\frac{\partial^2}{\partial x^2} + 2hk\frac{\partial^2}{\partial x\partial y} + k^2\frac{\partial^2}{\partial y^2}\right)f(p) + \cdots$$

The Hessian matrix of f is the matrix of the quadratic form in

$$H_pf = \begin{bmatrix} f_{xx} & f_{xy} \\ f_{yx} & f_{yy} \end{bmatrix}$$

The principal curvatures κ_1, κ_2 of the surface at p are the eigen values of H_pf [51] which are reals because the Hessian is a symmetric matrix. The mean curvature **H** of the surface at p is the trace $tr(H_pf)$ of H_pf

$$\mathbf{H} = f_{xx} + f_{yy} = \kappa_1 + \kappa_2$$

The Gaussian curvature **K** of the surface at p is the determinant of H_pf

$$\mathbf{K} = f_{xx}f_{yy} - f_{xy}^2$$

Example 5 A more general version of the minimal surface problem is given in terms of the mean curvature of a surface when $\mathbf{H} = 0$. The goal is to find the surface of smallest area bounded by a given closed curve in space. If the boundary curve can be projected down to a closed curve C surrounding a region Ω in the xy-plane and we express the surfaces $z = u(x, y)$, then the problem is to minimize the surface area functional

$$J[u] = \iint_{\Omega} \sqrt{1 + u_x^2 + u_y^2}\, dy dx$$

Subject to the boundary condition that $u(x, y)$ must assume prescribed values on C. Euler equation for this functional is

$$\frac{\partial}{\partial x}\left(\frac{u_x}{\sqrt{1+u_x^2+u_y^2}}\right) + \frac{\partial}{\partial y}\left(\frac{u_y}{\sqrt{1+u_x^2+u_y^2}}\right) = 0$$

which can be derived to $u_{xx}(1+u_y^2) - 2u_x u_y u_{xy} + u_{yy}(1+u_x^2) = 0$. This is the differential equation of minimal surfaces discovered by Lagrange. Euler showed that every nonplanar minimal surface must be saddle-shaped and also that $\mathbf{H} = 0$ at every point.

Example 6 (Minimal surface). $z = x^3 - 2xy^2$ (Fig. 1.2) is an example of minimal surface, this can be verified, replacing in the above differential equation. The graph is obtained by the command
>> ezsurf('x^3–2*x*y^2')

The concept of minimal surface establish a useful relation between variational principles and their applications, in particular the problems of computer vision. Several authors have studied these applications [43, 62] and there exists textbooks for a wider reading of the subject [21, 26].

Chapter 4
Interpolation. From One to Several Variables

We can use variational properties as a thread that conducts from univariate to multivariate approximations. We begin the chapter illustrating some limitations of the classical Lagrange interpolation and the way splines may help to a better formulation of the problem and the concrete algorithm to solve it. Here the reader will see how to apply the abstract theory of function spaces, both in the problem of curve reconstruction and its generalization to n dimensional data.

4.1 Introduction

Approximation theory can be seen as the art of representing a complex function in terms of a simpler one, in order to be implemented in numerical form, usually into a computer. Functions are abstract and ideal objects that appear taking part of infinite dimensional spaces, nevertheless algorithms must deal with finite linear combinations of functions, this implies that in any representation of a function there will be an intrinsic error that should be minimized.

Interpolation use ideas of approximation theory for modeling data. Given a set of data $D = \{(x_i, y_i)\}_{i=0}^{N}$, it is assumed that there exist a functional relation f between variables x and y such that $y_i = f(x_i)$. In the most simple case D represents two scalar variables whose values are obtained by sampling or experimentation; the nodes $a \leq x_0 < x_1 < \cdots < x_N \leq b$ belong to an independent variable x and y_i depends on x_i. The purpose of interpolation is to model this dependence making possible to estimate f at values that do not appear in data.

The traditional and simplest method for solving the interpolation problem in one variable is to build a polynomial $P_N(x) = \sum_{i=0}^{N} \alpha_i x^i$ of degree at most N, such that $y_i = P_N(x_i)$.

Polynomials are so frequently used in computational sciences that we can assume everybody is aware of their multiple applications. Nevertheless many readers may ask what is so important about polynomials. An important reason for considering the class of polynomials in the approximation of functions is that they are the

H. Montegranario, J. Espinosa, *Variational Regularization of 3D Data*, 31
SpringerBriefs in Computer Science, DOI 10.1007/978-1-4939-0533-1_4,
© The Author(s) 2014

Table 4.1 Divided differences

$x_0 \quad f[x_0]$

$$f[x_0, x_1] = \frac{f[x_1] - f[x_0]}{x_1 - x_0}$$

$x_1 \quad f[x_1]$ $f[x_0, x_1, x_2] = \dfrac{f[x_1, x_2] - f[x_0, x_1]}{x_2 - x_0} \cdots$

$$f[x_1, x_2] = \frac{f[x_2] - f[x_1]}{x_2 - x_1}$$

$x_2 \quad f[x_2]$

$\vdots$

simplest functions, from the point of view of its algebraic expression and representation in computers; operations on polynomials, such as derivative and indefinite
integrals are easy to calculate and happen to be also polynomials.

On the other hand many non-polynomial functions can be represented by power series that also reduce to polynomials. These facts are reinforced by Weirstrass
approximation theorem: *For any function $f(x)$ continuous in $[a,b]$, there exists a
sequence of ordinary polynomials which converges uniformly to $f(x)$ on $[a,b]$.*

Theorem 1 If $x_0, x_1, \ldots, x_N$ are $N+1$ distinct numbers and $y = f(x)$ a function with
$y_j = f(x_j)$, then there exists a unique polynomial of degree at most N such that
$y_j = P_N(x_j)$ for $j = 0,1,\ldots,N$. $P_N(x)$ is the classical Lagrange polynomial

$$P_N(x) = \sum_{k=0}^{N} y_k L_k(x), \text{with } L_k(x) = \prod_{\substack{j=0 \\ j \neq k}}^{N} \frac{(x - x_j)}{(x_k - x_j)} \tag{3.1}$$

For computational purpose is more convenient to use Newton's Divided-Difference
form for Lagrange polynomial written as

$$P_N(x) = f[x_0] + \sum_{k=1}^{N} f[x_0, x_1, \ldots, x_k](x - x_0) \cdots (x - x_{k-1}) \tag{3.2}$$

where the $f[x_0, x_1, \ldots, x_k]$'s are obtained by the well-known divided differences
method (Table 4.1)

$$\tag{3.3}$$

4.2   The Limitations of Polynomial Approximation

By Weirstrass approximation theorem the quality of polynomial interpolation can be measured with

$$\left\| f - P_N \right\|_\infty = \max_{a \leq x \leq b} \left| f(x) - P_N(x) \right|$$

if the function f is continuous on a,b one expect that P_N converge to f uniformly, that is $\left\| f - P_N \right\|_\infty \to 0$ as N goes to infinity. However it can be proved [33] that for any prescribed set of nodes $\{x_i\}$ there exists a continuous function f on $[a,b]$ such that the interpolating polynomials for f using these nodes fail to converge uniformly to f. One example for showing these difficulties is Runge's function $f(x) = 1/(1+x^2)$.

A common believe is that the larger the number of data points, the better the approximation. But while N increases P_N also increases the oscillations, then although P_N interpolates the N points, it also has a great distortion around these points.

The misbehavior of classical interpolation shows that more information is necessary if we want to obtain interpolants closer to the function generating data. The spline functions we will study in next section remedies this deficiency of Lagrange polynomial through a piecewise polynomial interpolation of low degree.

4.3   Spline Theory

Splines arise as a consequence of the lack of flexibility of high degree polynomials for interpolation. A spline function is also built from polynomials but they are now piecewise and built in such a manner they preserve the usefulness of polynomials.

Definition 2 Let $a = x_0 < x_1 < \cdots < x_N = b$ be a subdivision of the interval $[a,b]$ and $m \in \mathbb{N}$. A function $S : [a,b] \to \mathbb{R}$ is called a spline of degree m with respect to this subdivision if

1. On each subinterval $[x_i, x_{i+1}]$ S is a polynomial of degree $\leq m$
2. $S, S', S'', ..., S^{(m-1)}$ are all continuous functions on $[a, b]$.

This definition pose the question about what should be, under certain criteria, the most appropriate spline in terms of its degree. The answer is given by the variational interpretation of cubic polynomial spline. In order to minimize the oscillation produced by Lagrange polynomials of high degree, a good idea could be to minimize the integral $J = \int_a^b \kappa^2(x)\,dx$. Given that the curvature in Cartesian coordinates is $\kappa = y'' / (1 + y'^2)^{3/2}$, if the squared derivative in the denominator is small compared to 1 then $J[y] = \int_a^b (y'')^2\,dx$ is a good approximation. Assuming $y(x)$ sufficiently differentiable, calculating its first variation and applying integration by parts

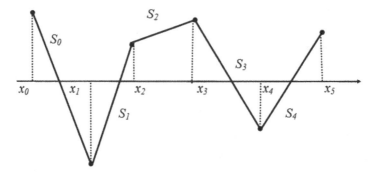

Fig. 4.1 A spline of degree 1

$$J[y+\varepsilon\varphi]=\int(y+\varepsilon\varphi)''^2\,dx$$

$$=\int y''^2+2\varepsilon y''\varphi''+\varepsilon^2\varphi''^2$$

$$\left.\frac{\partial J[y+\varepsilon\varphi]}{\partial\varepsilon}\right|_{\varepsilon=0}=\int 2y''\varphi''dx=0$$

$$=-(-)\int y^{IV}\varphi=0,$$

then by the fundamental lemma $y^{IV}(x)=0$. Solving this differential equations we obtain a cubic polynomial of the form $y(x)=a_3x^3+a_2x^2+a_1x+a_0$.

This result shows that the degree 3 in the interpolant polynomial as an optimal choice, in the sense that minimize a measure of the oscillations of the curve. Theorem 1 is a demonstration of this fact.

Definition 3 (Cubic spline interpolant) Let $a=x_0<x_1<\cdots<x_N=b$ be a subdivision of the interval $[a,b]$ and $D=\left\{(x_j,y_j)\right\}_{j=0}^N$ are $N+1$ points. The function $S(x)$ is called a cubic spline if there exist N cubic polynomials $S_j(x)$ with coefficients a_j,b_j,c_j,d_j such that

1. $S_j(x)=a_j+b_j\left(x-x_j\right)+c_j\left(x-x_j\right)^2+d_j\left(x-x_j\right)^3,\;x\in\left[x_j,x_{j+1}\right],\,j=0,1,\ldots,N-1.$

2. $S(x_j)=y_j$ $\qquad\qquad\qquad j=0,1,\ldots,N$

3. $S_j(x_{j+1})=S_{j+1}(x_{j+1})$ $\qquad j=0,1,\ldots,N-2$

4. $S'_j(x_{j+1})=S'_{j+1}(x_{j+1})$ $\qquad j=0,1,\ldots,N-2$

5. $S''_j(x_{j+1})=S''_{j+1}(x_{j+1})$ $\quad j=0,1,\ldots,N-2$ $\qquad\qquad$ (3.4)

These five conditions are not enough for assure uniqueness, so two more conditions must be imposed: If $S''(x_0)=S''(x_N)=0$, S is called **natural spline** and if $S'(x_0)=f'(x_0)$ and $S'(x_N)=f'(x_N)$ S is called **clamped spline**. In this way the number of conditions equals the amount of coefficients for the spline. On each of the N subintervals there is a different cubic polynomial $S_j(x)$ with four coefficients then we have to find $4N$ coefficients. Each $S_j(x)$ must satisfy two interpolation conditions on the ends of its corresponding interval, this gives $2N$ conditions. $S'_j(x)$ and $S''_j(x)$ must be continuous in the $N-1$ interior points which gives $2(N-1)$ conditions and finally for the natural spline second derivatives must vanish a at 2 points. We finally have $2N + 2(N-1) + 2 = 4N$ conditions equals to the number of coefficients.

Theorem 1 *Minimum property of cubic splines.* Assuming $f \in C^2[a,b]$ and $S(x)$ the unique interpolant natural cubic spline for $f(x)$ such that $S''(x_0) = S''(x_N) = 0$ then

$$\int_a^b [S''(x)]^2 \, dx \le \int_a^b [f''(x)]^2 \, dx$$

Proof. It's clear that

$$0 \le \int_a^b (f'' - S'')^2 \, dx = \int_a^b (f''^2 - 2f''S'' + S''^2) \, dx. \tag{3.5}$$

Now we calculate the second term $\int_a^b f''S'' \, dx$ in the former integral. Using integration by parts and the natural boundary condition $S''(x_0) = S''(x_N) = 0$ in (3.4) we can see

$$\int_a^b S''(f'' - S'') \, dx = S''(x)(f'(x) - S'(x)) \big]_a^b - \int_a^b (f' - S') S''' \, dx$$

$$= 0 - \int_a^b (f' - S') S''' \, dx$$

On every interval $[x_i, x_{i+1}]$ this last integral is zero because $S'''_j(x) = 6d_j$ and

$$\int_{x_i}^{x_{i+1}} (f' - S') S''' \, dx = 6d_i (f' - S') \big]_{x_i}^{x_{i+1}} = 0,$$

thus $\int_a^b S''(f''(x) - S''(x))(x) \, dx = 0$ and as a consequence $\int_a^b f''S'' \, dx = \int_a^b S''^2 \, dx$, and replacing in (3.4) $0 \le \int_a^b (f'' - S'')^2 \, dx = \int_a^b (f''^2 - S''^2) \, dx$, from which we finally obtain the expected result.

Up to now we have treated the spline in its piecewise definition, but sometimes we may need a more compact representation. This can be done by inserting splines into the formal language of linear spaces, finding nice, well-understood and easy to handle sets of basis functions $U = \{u_1, u_2, \ldots u_N\}$ to approximate f by linear combinations $\sum_{j=1}^N \alpha_j u_j$ and then measuring the error in the approximations. This implies that we need at least the structure of a linear space with a norm (normed space). In this kind of ideas there are results that provide the possibility to go from the piecewise definition of splines, convenient in computer representation, to more compact forms that are also useful for theory. The elements taken as basis functions are

$$(x-t)_+^n = \begin{cases} (x-t)^n, & x \geq t \\ 0, & x < t \end{cases}$$

Its well-known that $1, x, x, \ldots, x^k, (x-x_0)_+^k, (x-x_1)_+^k, \ldots, (x-x_{N-1})_+^k$ form a basis for the space of splines of order k [33]. Thus every cubic spline has a representation of the form

$$S(x) = \sum_{j=1}^{N} \alpha(x-x_j)_+^3 + \beta_1 + \beta_2 x, \tag{3.6}$$

and given the set $a = x_0 < x_1 < \cdots < x_N = b$ there is a unique natural cubic spline defined on these knots.

The form (3.6) is not well conditioned for computatonal work, however can be written in a more compact expression using translates of the form $|x - x_i|^3$, as in the next result.

Theorem 2 Given $a = x_1 < x_2 < \cdots < x_N = b$, every natural cubic interpolating spline $S(x)$ has a representation of the form

$$S(x) = \sum_{i=1}^{N} \alpha_i |x - x_i|^3 + \beta_1 + \beta_2 x, \, x \in \mathbb{R} \tag{3.7}$$

The coefficients $\{\alpha_j\}$ have to satisfy $\sum_{j=1}^{N} \alpha_j = \sum_{j=1}^{N} \alpha_j x_j = 0$ \tag{3.7b}

Given the points $D = \{(x_j, y_j)\}_{j=1}^{N}$, in order to determine the α_i's is necessary to solve de system

$$\begin{bmatrix} |x_1 - x_1|^3 & |x_1 - x_2|^3 & \cdots & |x_1 - x_N|^3 & 1 & x_1 \\ |x_2 - x_1|^3 & |x_2 - x_2|^3 & \cdots & |x_2 - x_N|^3 & 1 & x_2 \\ \vdots & \vdots & \ddots & \vdots & \vdots & \vdots \\ |x_N - x_1|^3 & |x_N - x_2|^3 & \cdots & |x_N - x_N|^3 & 1 & x_N \\ 1 & 1 & \cdots & 1 & 0 & 0 \\ x_1 & x_2 & \cdots & x_N & 0 & 0 \end{bmatrix} \begin{bmatrix} \alpha_1 \\ \alpha_2 \\ \vdots \\ \alpha_N \\ \beta_1 \\ \beta_2 \end{bmatrix} = \begin{bmatrix} y_1 \\ y_2 \\ \vdots \\ y_N \\ 0 \\ 0 \end{bmatrix}, \tag{3.8}$$

where the rows $1, \ldots, N$ in the above matrix correspond to the interpolation conditions in (3.7) and the last two rows are the application of (3.7b). The matrix is called **interpolation matrix**. If we define

$$P^t = \begin{bmatrix} 1 & 1 & \cdots & 1 \\ x_1 & x_2 & \cdots & x_N \end{bmatrix}, \; A_{ij} = |x_i - x_j|^3,$$

the system can be written as

$$\begin{bmatrix} A & P \\ P^t & 0 \end{bmatrix} \begin{bmatrix} \alpha \\ \beta \end{bmatrix} = \begin{bmatrix} y \\ 0 \end{bmatrix}, \tag{3.9}$$

with $\alpha = (\alpha_1,...,\alpha_N)^t$, $\beta = (\beta_1,\beta_2)^t$, $y = (y_1,...,y_N)^t$.

This result is demonstrated in the following section as a particular case of smoothing splines. By the moment in the next example we illustrate the power of this representation and its accuracy.

Example Comparing Lagrange polynomials with cubic splines. For this example, we took 10 equally spaced points $\{x_i\}_{i=1}^{10}$ in the interval $[-3,3]$. The y_i values correspond to evaluation of Runge's function $f(x) = 1/(1+x^2)$ on these points. The Lagrange polynomial is found in Newton's form (3.2) and the divided differences table (3.3). These are the matlab programs fnewton.m, newpol.m. The cubic spline is found by implementing (3.7) and (3.7b) and solving the linear system (3.9). The scalar or weights α and β are stored in the vector alfa of the function splinecubico.m that solve this system. The spline is evaluated with splcubic.m. The results are shown in Fig. 4.2. The oscillations of Lagrange polynomial and the performance of splines are evident. This means large data sets are better dealt with by splines than by Lagrange polynomials. In contrast to polynomials, the accuracy of the interpolation process using splines is not based on the polynomial degree but on the spacing of the data nodes.

4.4 Regularization and Smoothing Splines

In many real problems, data does not appear in exact form, frequently they are noisy. Noise is usually considered as an undesired perturbation and appears during every data acquisition process. In such case, it is not advisable to interpolate data $(y_i = f(x_i))$ but to approximate them $(y_i \approx f(x_i))$. Thus is better to apply the so called smoothing splines, which unlike interpolating splines, may not contain the data points (x_i, y_i). This problem is better solved by approaching the situation as an inverse problem that can be solved by minimization of the functional

$$J[f] = \sum_{j=1}^{N} (f(x_j) - y_j)^2 + \lambda \int f''^2 \, dx. \tag{3.10}$$

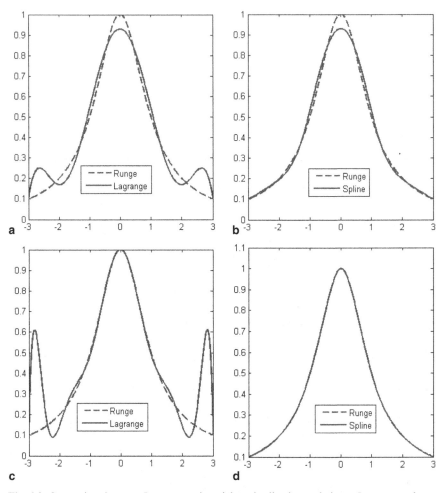

Fig. 4.2 Comparison between Lagrange polynmials and spline interpolation. **a** Lagrange polynomial (7 points). **b** Spline (7 points). **c** Lagrange (10 points). **d** Spline (10 points)

whose solution is exactly a natural cubic spline. The cost functional in (3.10)
has two parts: $E[f] = \sum_{j=1}^{N}(f(x_j) - y_j)^2$, that is minimized by a straight line and
$R[f] = \int f''^2(x)dx$, called **regularization functional,** minimized by a cubic spline.

The parameter λ trades off the importance of these two competing costs in
(3.10). For small λ, the minimizer is close to an interpolating spline. For λ large,
the minimizer is closer to a straight line. In statistics this method is sometimes
called penalized squares and it is seen as a form of nonparametric regression; a form
of regression analysis in which the predictor does not take a predetermined form
but it is constructed according to information derived from the data; its goal is to
construct a model for f and estimate it based on noisy data. In the case when both

the dependent variable y and the independent variable x are scalar variables, (3.10) relates dependent and independent variables as

$$y_i = f(x_i) + \varepsilon_i, i = 1,\ldots, N$$

where f is the regression function and ε_i are zero-mean independent random errors with common variance σ^2.

Theorem 3 The solution to the optimization problem (3.10) is a natural cubic spline which has a representation of the form

$$S(x) = \sum_{j=1}^{N} \alpha_j \phi(|x - x_j|) + p(x), \ x \in \mathbb{R}$$

where $\phi(r) = r^3, r \geq 0$ and $p \in \Pi_1(\mathbb{R})$. The coefficients $\{\alpha_j\}$ have to satisfy (3.7b). In the same way, for every set $A = \{x_1, x_2, \ldots, x_N\} \subseteq \mathbb{R}$ of pairwise dintict points and for every $y \in \mathbb{R}^N$, there exists a function S of the form (3.7) with (3.7b) that interpolates the data, this is $S(x_j) = y_j, 1 \leq j \leq N$.

Proof

Applying the properties of Dirac's distribution, the functional (3.10) can be written as

$$J[f] = \sum_{j=1}^{N} (f(x_j) - y_j)^2 + \lambda \int f''^2 dx = \sum_{j=1}^{N} \int (f(x) - y_j)^2 \delta(x - x_j) + \lambda \int_a^b f''^2 dx$$

$$= \int \sum_{j=1}^{N} (f(x) - y_j)^2 \delta(x - x_j) + \lambda \int f''^2 dx$$

Calculating the first variation of the functional J[f]

$$\frac{\delta J}{\delta f} = 2 \sum_{j=1}^{N} \left(f(x) - y_j \right) \delta(x - x_j) + \lambda \frac{d^4 f}{dx^4} = 0$$

$$\frac{d^4 f}{dx^4} = \frac{1}{\lambda} \sum_{j=1}^{N} (y_j - f(x)) \delta(x - x_j)$$

$$\frac{d^4 f}{dx^4} = \sum_{j=1}^{N} \frac{(y_j - f(x))}{\lambda} \delta(x - x_j)$$

The solution to this differential equation can be written as the convolution of the source term with the fundamental solution of the operator $f^{IV} = \delta$. So we calculate the Green's function $K(x, \xi)$ such that $\dfrac{d^4 K(x, \xi)}{dx^4} = \delta(x - \xi)$, then $K(x, \xi) = |x - \xi|^3$, and

$$f(x) = K * \sum_{i=1}^{N} \frac{y_i - f(x)}{\lambda} \delta(x - x_i)$$

Defining $\alpha_i = \dfrac{y_i - f(x_i)}{\lambda}$ (3.11)

we have the spline solution

$$f(x) = \sum_i \alpha_i |x - x_i|^3$$

As the null space of R[f] is $\Pi_1(\mathbb{R})$, the polynomials of degree less or equal than one, we have to add a term $p(x) \in \Pi_1(\mathbb{R})$ written as a linear combination of any base. For $\Pi_1(\mathbb{R})$ in particular we can use $\{1, x\}$. The final expression for the solution of the optimization problem is

$$S_\lambda(x) = \sum_i \alpha_i \phi(|x - x_i|) + p(x) = \sum_i \alpha_i |x - x_i|^3 + \beta_1 + \beta_2 x \qquad (3.12)$$

In order to find the values for $\{\alpha_i\}$ we translate the problem to the matrix form. Using (3.11)

$$y_i = f(x_i) + \lambda \alpha_i \text{ then } y_i = \sum_{j=1}^{N} \alpha_j |x_i - x_j|^3 + \beta_1 + \beta_2 x_i + \lambda \alpha_i, i = 1, ..., N.$$

we obtain $y = A\alpha + P\beta + \lambda\alpha$ and finally the system

$$\begin{cases} (A + \lambda I)\alpha + P\beta = y \\ P'\alpha = 0 \end{cases}$$

and in matrix form

$$\begin{bmatrix} A + \lambda I & P \\ P' & 0 \end{bmatrix} \begin{bmatrix} \alpha \\ \beta \end{bmatrix} = \begin{bmatrix} y \\ 0 \end{bmatrix}. \qquad (3.12b)$$

This is a very practical result because although the problem (3.10) is much more complex than interpolation without noisy data shown in (3.9), the only difference is the term λ added to the diagonal of the matrix A.

4.5   Extensions to Higher Dimensions

Historically a great interest in application and generalizations of these techniques arouse after the application of variational approach to splines theory, especially their extensions to several variables. In fact, cubic spline is a particular case of the problem considered by Schoenberg, about finding and interpolator that minimize the functional $\int_a^b (f^{(m)}(x))^2\,dx$ in the Sobolev space W^m of functions of $m-1$ continuous derivatives and m-th derivative square integrable.

During the 1970s Wahba and others demonstrated the connection between smoothing splines and Bayes estimates [70, 71], showing their good approximation and theoretical properties. In this framework f may be a fixed function with certain degree of smoothnes or f can be considered as a sample function from a stochastic process [68]. Thanks to the increase in computers capacity, these results became feasible and attractive for many applications.

The evolution of interpolation and approximation methods from one to several variables has taken many forms: tensor products, piecewise bivariate polynomials and box splines, just to mention the most important. Some of them aim to reproduce in further dimensions the properties of cubic splines. Nevertheless the most successful idea to extend splines and smoothing splines to several variables has been by generalizing its variational characterization, obtained by Duchon, Meinguet and others in the form of thin plate splines (TPS). Besides its theoretical properties and physical interpretations, TPS has representation as radial basis function (RBF) which gave it success in practical implementations.

A RBF is a linear combination of translates of a fixed radial function $\Phi = \phi(\|\cdot\|)$. The function is called radial because it is the composition of a univariate function with the Euclidian norm on $\mathbb{R}$. In this sense it generalizes to $\mathbb{R}^n$, where the name radial is more 'natural'. A straightforward generalization is to build interpolants of the form

$$S(x) = \sum_{i=1}^{N} \alpha_i \phi(\| x - x_i \|_2) + p(x),\ x \in \mathbb{R}^n \tag{3.13}$$

Where $\phi : [0,\infty) \to \mathbb{R}$ is a univariate fixed function and $p \in \Pi_{m-1}(\mathbb{R}^n)$ is a low degree n-variate polynomial. The additional conditions on the coefficients now become

$$\sum_{j=1}^{N} \alpha_j q(x_j) = 0 \text{ for all } q \in \Pi_{m-1}(\mathbb{R}^n) \tag{3.13b}$$

In many cases it is not necessary the additional polynomial $p(x)$, so that one does not need the additional conditions (3.13b). In this particular case the interpolation problem reduce to the question whether the matrix $A = (\phi(\| x_i - x_j \|))$ is nonsingular. In other words, does there exists a function $\phi : [0,\infty) \to \mathbb{R}$ such that for all pairwise distinct points $x_1,...,x_N \in \mathbb{R}^n$ the matrix A is nonsingular

The answer is yes. Examples are the Gaussian $\phi(r) = e^{-cr^2}$ $c > 0$ and the multiquadric $\phi(r) = \sqrt{c^2 + r^2}$ $c > 0$. In the case of Gaussian function the matrix is also positive definite, although with some restrictions, this can be true for functions ϕ with compact support. The extension to further dimensions is the subject of the following chapters.

4.6   Matlab Codes for Chapter 4

```
%Call as [c,d]=newpol(x,y)
%x,y: row data vectors
%d:divided differences matrix
%c:diagonal of d  for f[xo..xN] in Newton form
function [c,d]=newpol(x,y)
N =length(x);
d=zeros(N,N);
d(:,1)= y';

for j=2:N
    for k=j:N
        d(k,j)=(d(k,j-1)-d(k-1,j-1))/(x(k)-x(k-j+1));
    end
end
c=diag(d);


%-----------------------------------------------------

function w = fnewton2(c,xdata,x)
%call as w = fnewton(c,xdata,x)

%Evaluate Newton's Lagrange form
% x: evalution
% c diagonal of divided differences matrix
%xdata : knots
N= length(xdata);
M=length(x);
w=zeros(M,1);
for j=1:M
w(j) =c'*cumprod([1;x(j).*ones(N-1,1)-xdata(1:N-1)]);
end


%-----------------------------------------------------

function A =  matrizcubica(td)
%td: x-coordinates of the knots
M =length(td);

unos=ones(M,1);
A1 = td*unos';
B =(abs(A1-A1')).^3;

P =[unos td];
A= [ B P;P' [0 0;0 0]];

%-----------------------------------------------------
```

```
function val= splcubic(alfa,td,x)
%Evaluate spline in x
%alfa: column with the weights of the spline
%td column of knots
%x evaluation vector

M=length(x);
N=length(td);
ax = x'*ones(1,N);
at=ones(M,1)*td';
A = abs(ax-at).^3;

val = A*alfa(1:N)+alfa(N+1)*ones(M,1)+alfa(N+2)*x';

%----------------------------------------------------
function alfa=splinecubico(td,y)
%Obtains the weights for spline
%td x-coordinates of knots
% y: data to be interpolated
% alfa:weight vector
A = matrizcubica(td);
b= [y;0 ;0];
alfa= A\b;

%----------------------------------------------------

function comparar(a,N)
%Draw graphs for Runge,Lagrange and spline functions
 clc;close all;

g=@(x) 1./(1+x.^2);
xknots=(-a:2*a/N:a)';
y=g(xknots);
[c,d]=newpol(xknots,y);
t=-a:0.01:a;
y3=g(t);                        %Runge
y1=fnewton2(c,xknots,t);        %Lagrange
L=splinecubico(xknots,y);
y2=splcubic(L,xknots,t);        %cubic spline

figure
plot(t,y3,'--','LineWidth',2)%Runge +Lagrange
hold on
plot(t,y1,'LineWidth',2)
figure
plot(t,y3,'--','LineWidth',2)%Runge+spline
hold on
plot(t,y2,'LineWidth',2)

Example:
>>comparar(3,10)
```

Chapter 5
Functionals and Their Physical Interpretations

As we saw in the previous chapter, the variational interpretation of splines, suggests to apply physical or geometrical references to other problems of reconstruction from data. It is possible to go deeper into this method and obtain other models in the form of functionals that could be added or included in the cost function of variational regularization. This time not only for one dimensional data, but for further dimensions. In order to do a heuristic discussion we consider the behavior of curves and surfaces under geometrical criteria, as for instance, Gaussian and mean curvature as well as physical criteria, based on the beam and plate models of continuum mechanics (Fig. 5.1).

5.1 Physical Interpretations of Functionals

The curves and surfaces generated by elastic media are physical models that provide attractive metaphors and interpretations of smoothness: strings, beams, membranes, and plates (Fig. 5.1, 5.2). With these physical models it is not difficult to visualize the properties of surfaces restrained around their boundaries, and deflected by lateral forces applied perpendicular to the rest positions.

An elastic body is defined to be a solid which, if deformed, will revert to its original shape once the forces causing the deformation are removed. The problems of stable equilibrium, are governed by the variational principle of minimal potential energy. The potential energy is the work done by deformation forces as a function of the shape of the deformed surface. The idea of smoothness and constraints can be represented by the laws of physical elastic surfaces and potential energies are functionals represented by integrals extended to space regions, surfaces, or lines. The variational principles can be established in differential form as boundary valued problems by applying Euler-Lagrange equations. Thus, the integral and differential formulations provides different insights of the problem.

H. Montegranario, J. Espinosa, *Variational Regularization of 3D Data*,
SpringerBriefs in Computer Science, DOI 10.1007/978-1-4939-0533-1_5,
© The Author(s) 2014

Fig. 5.1 A real example of Dirac functionals in continuum mechanics (La Ferté pedestrian bridge in Stuttgart, Germany). (Source, Google maps)

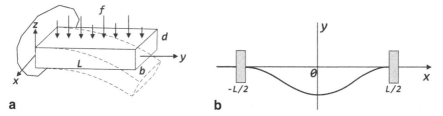

a   b

Fig. 5.2 Beams are a useful metaphor for splines physical interpretations

5.2 Beams and Splines

The beam model is used as physical representation of splines. Strings and beams are the simplest examples of continuum mechanics phenomena, and can be considered as one dimensional. Their 3D equivalent are membrane and plates. Using the arguments of potential energy it is possible to show that beams and plates are more convenient in the physical interpretations of splines [62].

A beam (Fig. 5.2a) is a body rectilinear in shape and whose length L is considerably larger than its two other dimensions, width b and depth d. So the beam is modelled as a one-dimensional object (Fig. 5.2b)

In elasticity theory the following situation is studied: a cylindrical solid beam, elastic and homogeneous has its ends fixed at the same height above the ground, the problem is to determine the profile of the deflected axis of the beam. That is, we want to find the in-plane deflection u of the beam while subject to a load f. A fundamental assumption in this theory is that no deformation occur in the plane of the cross section of the beam. When the beam bends, it becomes curved, and the bending produces internal forces that try to straighten it. The restoring force is governed by the curvature and the displacement u is perpendicular to the beam.

The Euler-Bernoulli equation describes the relationship between the beam's deflection and the applied load f [64, 65]

$$\frac{d^2}{dx^2}\left(EI\frac{d^2u}{dx^2}\right) = f \tag{5.1}$$

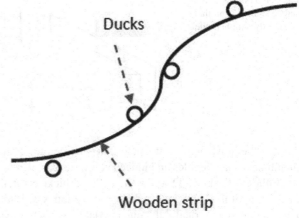

where E: modulus of elasticity, I: moment of inertia of a cross section about the neutral axis. The curve $u(x)$ describes the deflection of the beam in the vertical direction at some position x, f is force per unit length. Sometimes the product EI (known as the *stiffness*) is a constant, so that

$$EI \frac{d^4 u}{dx^4} = f.$$

(5.2)

The Euler-Bernoulli beam theory plays an important role in structural analysis. Depending of the assumptions, several beam theories have been developed.

The mechanical model of a cubic spline is shown in Fig. 5.3. The term spline is derived from the analogy to a draughtsman's approach to pass a thin metal or wooden strip through a given set of constrained points called ducks. These ducks are the physical equivalent of the interpolation points. Because the beam is unloaded (the load $f = 0$) between the ducks, solving equation $d^4 u / dx^4 = 0$, each segment $u(x)$ of the spline curve is a cubic polynomial. At the ducks, there exist a concentrated load of magnitude say u_0, then

$$\frac{d^4 u}{dx^4} = u_0 \delta(x),$$

this implies $u'''(x)$ is a step function and then the second derivative $u''(x)$ is continuous. The second derivative of the spline is zero at the end points because there is no bending moment there. The resulting curve is known as *the natural cubic spline* since these end conditions occur naturally in the beam model.

5.3   Plate and Membrane Models

Plates and membrane models appear in everyday situations, for example, a plate may be part of a condenser microphone, such as one may find inside a telephone or a cellular phone. When the user speaks, the acoustic pressure produced by sound waves

Fig. 5.4 Thin plate spline can be studied from this apparently simple model

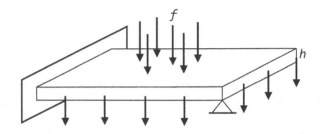

induces the movement of the plate. The surface of a drum is a classic example of a membrane, soap bubbles and indeed cell walls can also be modeled as membranes.

A **plate** is the body $\Omega \times]-h/2, h/2[$ of thickness h, that occupies the region Ω in $\mathbb{R}^2$, one of its dimensions, in the z direction, say, is much smaller than the other two, so that geometrically the plate is flat (Fig. 5.4). A plate theory takes advantage of this disparity in length scale to reduce the full three-dimensional solid mechanics problem to a two-dimensional problem. The aim of plate theory is to calculate the deformation and stresses in a plate subjected to loads.

Membranes can be considered as extremely thin plates with no flexural rigidity. The structural behavior of stretched membranes resembles that of a network of stretched springs. Nevertheless membranes may form surfaces with discontinuous first derivatives. In contrast, thin plate surfaces do have flexural rigidity; hence, they are generally a smoother class of surfaces than membranes surfaces.

Now, suppose that the membrane at rest covers a region Ω of the plane and that the deformation $u(x, y)$ is normal to the equilibrium plane. Suppose this deformation is small in the sense that higher powers of the partial derivatives u_x, u_y of u are negligible compared with lower ones. Then the expression $\iint_\Omega \sqrt{1 + u_x^2 + u_y^2}\, dy\, dx$ for the area may be replaced by $\iint_\Omega [1 + \frac{1}{2}(u_x^2 + u_y^2)]\, dy\, dx$ and the potential energy is proportional to the functional

$$U[u] = \iint_\Omega \left(u_x^2 + u_y^2\right) dy\, dx \tag{5.3}$$

For the equilibrium problem of the membrane we suppose the displacement $u(x, y)$ of the membrane have prescribed values on its boundary and that no external forces act on the membrane.

The displacement $u(x, y)$ in the equilibrium position is the function for which the potential energy functional (5.3), attains the least possible value among the functions which are continuous in the closed set Ω, take on the prescribed boundary value, and possesses continuous first and piecewise continuous second derivatives in the interior. The Euler equation of the functional is

$$\Delta = u_{xx} + u_{yy} = 0$$

Thus the problem of finding the equilibrium position is equivalent to the boundary value problem: find in Ω a solution of the above partial differential equation (the potential equation) which assume prescribed values on $\partial\Omega$.

The thin plate spline involves partial derivatives of higher order than membrane [31]. It is known that the potential energy of a thin plate under deformation is given in terms of the principal curvatures κ_1, κ_2 of the deformed plate's surface, the potential energy density is given by

$$\frac{A}{2}(\kappa_1^2 + \kappa_2^2) + B\kappa_1\kappa_2 = 2A\left(\frac{\kappa_1 + \kappa_2}{2}\right)^2 - (A - B)\kappa_1\kappa_2$$
$$= 2AH^2 - (A - B)K$$

H, K the median and Gaussian curvatures, A, B are constants dependent of the plate material. Assuming small deflection u and its partial derivatives, it can be shown

$$H = \kappa_1 + \kappa_2 \approx \Delta u, \text{ and } K = \kappa_1\kappa_2 \approx u_{xx}u_{yy} - u_{xy}^2$$

Thus, the potential energy density can be approximated as $(\Delta u)^2 - (1 - \sigma)\left(u_{xx}u_{yy} - u_{xy}^2\right)$ apart from the flexural rigidity A, which can be set to 1, and $\sigma = B/A$ is the Poisson's ratio which represents the change in width as the material is stretched lengthwise. The potential energy of the deformation is

$$J[u] = \int_\Omega \left[(\Delta u)^2 - (1 - \sigma)\left(u_{xx}u_{yy} - u_{xy}^2\right)\right]d\mathbf{x}$$

As in the case of beam model for cubic spline, if we consider a clamped plate with natural boundary conditions, the former expression reduce to the thin plate functional

$$J[u] = \int_\Omega \left[u_{xx}^2 + 2u_{xy}^2 + u_{yy}^2\right]dydx$$

Thus, the problem of surface reconstruction can be reinterpreted in terms of the plate model. The interpolation problem consists in determining the surface formed by fitting a thin elastic plate over a region and through the known points. A thin plate surface is better in some respects than a membrane surface. This is because, contrary to membrane, the plate has a nonzero flexural rigidity. By this property, the plate is able to extrapolate beyond peripheral constraints, something that a membrane cannot do.

Continuum mechanics, in particular elasticity theory, provide a great amount of real models for variational regularization. The models shown here are simple compared with others that take into account more complex behaviours of elastic materials; for more details the reader may consult the references [31, 64, 65].

Chapter 6
Regularization and Inverse Theory

Now we are going to apply inverse theory to formalize the techniques developed in former chapters, especially the regularization method used for cubic splines. Inverse theory is one of the great achievements of applied mathematics and its applications are at the core of modern technology. Thus we formulate the theory in general terms and then, to the interpolation problem.

6.1 Introduction

A proper computational formulation of reconstruction problems should include the fact they are inverse processes that are mathematically ill-posed. The concepts of regularization theory give a comprehensive framework for the formulation, algorithms and implementation of these problems. Furthermore, the mathematical theory of regularization provides a tool for incorporating prior knowledge, constraints, and quality of solutions. 3D-reconsruction can be regarded as the solution to an inverse problem in the sense of Hadamard. Thus, the theory developed for regularizing ill-posed problems leads in a natural way to this solution in terms of inverse theory.

While from the point of view of optics the problem is to determine the images of physical objects, computer vision is dealing with the inverse problem of recovering three-dimensional shape from the light distribution in the image. In surface reconstruction one is faced with the task of discovering the nature of objects that produced a set of points for the surface received by a camera or some other kind of acquisition process.

H. Montegranario, J. Espinosa, *Variational Regularization of 3D Data*,
SpringerBriefs in Computer Science, DOI 10.1007/978-1-4939-0533-1_6,
© The Author(s) 2014

6.2   Computational Mathematics in Operator Language

A large number of problems that appear in engineering and computational mathematics can be formulated in terms of operator's formalism as: Solve the equation

$$Au = v,$$ (6.1)

where $A : \mathcal{U} \to \mathcal{V}$ may be a linear or nonlinear operator over normed spaces $\mathcal{U}, \mathcal{V}$. To be more detailed, three distinct types of problems are recognized.

1. The **direct** problem. Given A and u find v. For example the computation of a definite integral
2. The **inverse** problem. Given A and v, find u. Examples are solving systems of simultaneous equations, ordinary and partial differential equations and integral equations
3. The **identification** problem. Given u and v, find A

In other words, let's suppose, u, v and A represent the **input**, **output** and the **system**, respectively. Thus in a direct problem, we need to determine the output of a given system generated by a known input; in the inverse problem one looks for the input which generates a known output. In the identification problem from a knowledge of the relation between input and output we want to find the laws governing a system. Usually only a finite number of input- output pairs is known.

From the point of view of modern mathematics, all problems can be classified as being either correctly posed or incorrectly posed. In the language of functional analysis, this fact can be stated in the following manner.

Let $\mathcal{U}$ and $\mathcal{V}$ be Banach spaces and the continuous operator $A : \mathcal{U} \to \mathcal{V}$ (not necessarily linear). The equation $Au = v$, represents a correct, correctly posed or **well-posed** problem in the sense of Hadamard, if the operator A has a continuous inverse $A^{-1} : \mathcal{V} \to \mathcal{U}$. In other words

(**wp1**) Exists a solution: $\forall v \in \mathcal{V}$ there exists a solution $u \in \mathcal{U}$.

(**wp2**) The solution is unique: $\forall v \in \mathcal{V}$ there is no more than one $u \in \mathcal{U}$ such that

$$Au = v$$

(**wp3**) The solution u^* depends continuously on the data: $\| u - u^* \|_{\mathcal{U}} \to 0$ when

$$\| v - v^* \|_{\mathcal{V}} \to 0 .$$

If one of these three conditions is not satisfied, the problem $Au = v$ is called **ill-posed**. Given that we want to find u given v it is also called **ill-posed inverse problem**.

These conditions do not have the same degree of importance [15]. If the uniqueness condition (**wp2**) is not satisfied then the problem does not make sense. However, if (**wp1**) is not satisfied, it only means that there are not conditions to guarantee existence of a solution. On the other side, one may think (as Hadamard did) that without (**wp3**) the problem $Au = v$ does not have physical sense and is incomputable. Nevertheless, choosing a proper notion of convergence and the space $\mathcal{U}$, it is possible to fix the situation. For instance $\mathcal{U}$ and $\mathcal{V}$ may be taken from the classical spaces $C^k(\mathbb{R}^n)$ or $H^m(\mathbb{R}^n)$. These spaces are a natural setting for mathematical physics and partial differential equations. They reflect physical reality and generate stable computational algorithms

During the last decades there has been a growing interest in some important problems which fail to have some or all of the defining properties of a well-posed problem [50, 63]. Many, but not all, are concerned with questions about the cause of a given effect. It should be noted that a problem is called an inverse one only because of its relation to another that we call direct. Also, not all inverse problems are ill-posed, nor are all ill-posed problems, inverse problems. It may be shown that many ill-posed and/or inverse problems may be reduced, perhaps after some linearization, to the operator equation $Au = v$.

A typical example of an ill-posed problem is that of a linear operator equation with A a **compact operator**, in this case both conditions for being Hadamard well-posed can be violated. For example, if $\mathcal{U}$ is infinite dimensional, then A^{-1} may not be defined in all of $\mathcal{V}$ ($A(\mathcal{U}) \neq \mathcal{V}$) and, secondly, A^{-1} (defined on $A(\mathcal{U}) \subset \mathcal{V}$) may not be continuous.

6.3 Compact Operators

The theory of inverse problems can be well described in terms of the theory of compact operators. From a practical point of view we would like to extend to infinite dimensional spaces, the good properties we find in the operations of linear algebra for the finite-dimensional case; in spite of the difficulties of the infinite dimensional case. Compact operators are a tool for maintaining a great part of the desirable properties of finite dimensional linear algebra, as for example spectral theory.

Definition 1 A bounded linear operator $A : \mathcal{U} \to \mathcal{V}$ is called **compact** if for each bounded set in B in $\mathcal{U}$, the set $A(B)$ has compact closure in $\mathcal{V}$.

The definition can be rephrased as follows. An operator A is compact if and only if for a given bounded sequence $\{x_j\} \subset \mathcal{U}$ the sequence $\{Ax_j\}$ has a convergent subsequence in $\mathcal{V}$. If the $\{x_j\}$ lie in a finite dimensional space, then this is true for every bounded operator.

Example Any linear operator $A : \mathcal{U} \to \mathcal{V}$ for which the range $\mathcal{R}(\mathcal{U})$ is finite dimensional is compact. As a consequence, matrix operators are compact

Example An example of compact operator where a, b are finite and $K(s, t)$ is continuous, is $A : \mathcal{L}^2[a, b] \to \mathcal{L}^2[c, d]$, defined by

$$(Ax)(s) = \int_a^b K(s, t) x(t) dt,$$

For linear operators, compactness is a stronger condition than boundedness; every compact operator is bounded, but not vice versa. The set of all compact operators $A : \mathcal{U} \to \mathcal{V}$ is a closed subspace of $\mathcal{L}(\mathcal{U}, \mathcal{V})$. This means that if $A, B : \mathcal{U} \to \mathcal{V}$ are compact operators then so are the operators $A + B$ and $\alpha A, \alpha \in \mathbb{R}$.

Definition 2 (Singular values) Let $A : \mathcal{U} \to \mathcal{V}$ be a compact operator on Hilbert spaces $\mathcal{U}$, $\mathcal{V}$ with adjoint operator $A^* : \mathcal{V} \to \mathcal{U}$. The square roots $\sigma_j = \sqrt{\lambda_j}$ of the eigenvalues λ_j of the self-adjoint operator $A^* A : \mathcal{U} \to \mathcal{U}$ are called **singular values** of A. every eigenvalue λ of $A^* A$ is nonnegative.

Theorem (Singular value decomposition) Let $A : \mathcal{U} \to \mathcal{V}$ be a compact operator, $A^* : \mathcal{V} \to \mathcal{U}$ its adjoint operator and $\sigma_1 \geq \sigma_2 \geq \sigma_3 \cdots > 0$, the ordered sequence of singular values of A. Then there exists two orthonormal systems $\{x_j\} \subset \mathcal{U}$ and $\{y_j\} \subset \mathcal{V}$ with the following properties:

$$Ax_j = \sigma_j y_j$$

$$A^* y_j = \sigma_j x_j$$

The system $\{x_j, y_j, \sigma_j\}$ is called a singular system for A. Every $x \in \mathcal{U}$ possesses the Singular Value Decomposition (SVD)

$$x = x_0 + \sum_{j \in J} (x, x_j) x_j$$

for some $x_0 \in \mathcal{N}(A)$ and

$$Ax = \sum_{j \in J} \sigma_j (x, x_j) y_j \tag{6.2}$$

A finite-dimensional version of singular value decomposition for matrices asserts that any $m \times n$ matrix A can be decomposed as

$$A = V \Sigma U^t \tag{6.3}$$

where V is $m \times m$ and orthogonal and U is $n \times n$ and orthogonal and Σ has the form

$$\Sigma = \begin{bmatrix} D & 0 \\ 0 & 0 \end{bmatrix} \qquad D = \begin{bmatrix} \sigma_1 & & \\ & \ddots & \\ & & \sigma_k \end{bmatrix}$$

If $A = V \Sigma U^t$ is the SVD of A then its pseudo-inverse A is defined as

$$A^\dagger = U \Sigma^\dagger V^t \tag{6.4}$$

where $\Sigma^\dagger$ is the $n \times m$ matrix

$$\Sigma^\dagger = \begin{bmatrix} D^{-1} & 0 \\ 0 & 0 \end{bmatrix}, \text{ with } D^{-1} = \begin{bmatrix} \sigma_1^{-1} & & \\ & \ddots & \\ & & \sigma_k^{-1} \end{bmatrix}$$

The pseudo-inverse gives the classical least square solution $\tilde{x}$ to the problem $\mathbf{Ax} = \mathbf{b}$ which minimize $\|Ax - b\|$ with $\|\tilde{x}\|$ minimum. $A^\dagger$ is called the Moore-Penrose inverse.

6.4 Regularization and Inverse Problems

Regularization theory provides a framework for formulation of inverse problems. A general formulation of inverse problems begins with the sensed data v, which is produced through the action of an operator A, acting upon the data, then the goal is to recover u from the operator equation $Au = v$. A problem involving inversion of this operator is well-posed [15] if we can ensure existence, uniqueness, and stability of solutions.

There may be many reasons for the failure of these conditions, making the problem ill-posed. For example, A may be invertible but ill-conditioned, i.e. small changes in data lead to large deviations in the solution, a situation that could be worse with noisy data. A may not have a unique solution, in this case we have to add more information in order to restrict the solution space, or A may be of rank greater than the number of degrees of freedom making the system over-determined and thus a solution may not exist if any of the measurements contains noise; regularization theory provides mathematical tools that enable one to turn an ill-posed problem into a well posed problem.

The main idea supporting regularization methods is that the solution of an ill-posed problem can be obtained using a variational principle, which contains both the data and prior smoothness information. For the stable approximation of a solution of the operator equation $Au = v$, it is assumed that only noisy data v_δ of the exact data v are available. Tikhonov or variational regularization aims at minimizing a relatively flexible model, and then controls the variance by modifying the error function by the addition of a penalty term so that the total error function becomes the improved functional

$$J_{\lambda,v_\delta} = \underbrace{E[Au,v_\delta]}_{\text{fidelity to data}} + \underbrace{\lambda R[u]}_{\text{smoothness}} \tag{6.5}$$

only assuming that E is a functional measuring the error between Au and v_δ, $\lambda > 0$ and $R[\cdot]$ is a nonnegative functional. The real number λ is called *regularization parameter*. The idea is to find a solution for $Au = v$ as an element minimizing the functional (6.5). One may think there could be easier solutions. For example, the solution to $min\|Au - v\|$ but, but it does not work because this is equivalent to $Au = v$

and therefore will also be ill-posed. Another idea could be to use Moore-Penrose generalized inverse $u^\dagger = A^\dagger v$, defined as the minimum norm solution of the problem $\min_{u \in \mathcal{U}} \| Au - v_\delta \|_{\mathcal{U}}^2$. However, the operator $A^\dagger v$ is usually not bounded.

The idea of regularization is to replace the generalized inverse $A^\dagger$ by a family of continuous operators A_λ, which converge point- wise to $A^\dagger$. The regularization parameter λ is chosen to depend on noise and data. Under certain conditions the regularized solution converges to $A^\dagger v$ as $\lambda \to 0$.

In general, let $\mathcal{U}$ and $\mathcal{V}$ be two Hilbert spaces and $A : \mathcal{U} \to \mathcal{V}$ a linear bounded operator. The Tikhonov regularization scheme [63] is given by

$$\min_{u \in \mathcal{U}} J_{\lambda, v_\delta}[u] = \| Au - v_\delta \|_{\mathcal{U}}^2 + \lambda \| u \|_{\mathcal{U}}^2 \tag{6.6}$$

The solution can be seen calculating the first variation of the functional $J[u]$

$$J_\lambda[u + \varepsilon \varphi] = \| A(u + \varepsilon \varphi) - v_\delta \|^2 + \lambda \| u + \varepsilon \varphi \|^2$$

$$= \| Au + \varepsilon A\varphi \|^2 - 2\langle Au + \varepsilon A\varphi, v_\delta \rangle + \| v_\delta \|^2 + \lambda \left(\| u \|^2 + 2\varepsilon \langle u, \varphi \rangle + \varepsilon^2 \| \varphi \|^2 \right)$$

$$\frac{\partial J_\lambda}{\partial \varepsilon} = 2\langle Au, A\varphi \rangle + 2\varepsilon \| A\varphi \|^2 - 2\langle A\varphi, v_\delta \rangle + 2\lambda \left(\langle u, v_\delta \rangle + \varepsilon \| \varphi \|^2 \right)$$

$$\frac{\partial J_\lambda}{\partial \varepsilon} \Big|_{\varepsilon=0} = \langle Au, A\varphi \rangle - \langle A\varphi, v_\delta \rangle + \lambda \langle u, v_\delta \rangle = 0$$

$$\langle Au - v_\delta, A\varphi \rangle + \lambda \langle u, v_\delta \rangle = 0,$$

using the properties of adjoint operators

$$\langle (A^* A + \lambda I)u - A^* v_\delta, \varphi \rangle = 0$$

$$(A^* A + \lambda I)u = A^* v_\delta,$$

Thus we obtain the solution

$$u_\lambda^\delta = (A^* A + \lambda I)^{-1} A^* v_\delta,$$

called the Tikhonov approximation to $A^\dagger v$. Since the inverse operator in the right hand side is bounded, the Tikhonov approximation u_λ depends continuously on y for each fixed $\lambda > 0$ This is the classical solution that can be also obtained by operator theory [15].

Using the SVD decomposition, it is possible to show a limiting process such that Tikhonov regularization, approximates the least squares solution $A^\dagger v$ that is, $u_\lambda \to A^\dagger v$ as $\lambda \to 0$ in the sense that $\lim_{\lambda \to 0} \|u_\lambda - A^\dagger v\|^2 = 0$. To put it another way, an ill-posed problem is approximated by a family of nearby well-posed problems.

6.5 Interpolation as an Inverse Problem

Now we give a formal interpretation of surface reconstruction in terms of inverse theory. The data to be interpolated, can be seen as evaluation functionals $L_{a_i}(f) = f(a_i)$ in a Hilbert function space $\mathcal{H}$. As these functionals are bounded, we can apply the Riesz representation theorem to say that for each x in $\Omega \subset \mathbb{R}^n$ there exists a unique representer $K_x \in \mathcal{H}$ such that $\langle f, K_x \rangle = f(x)$.

We say the mapping $K : \Omega \times \Omega \to \mathbb{C}$ such that $K(x, y) = K_x(y)$ is the **reproducing kernel** (or simply kernel) of $\mathcal{H}$. In a similar way to Dirac's delta functional, the reproducing kernel is so called because it has the potential of reproducing each function in $\mathcal{H}$: $\langle f, K_x \rangle = f(x)$ for all f in $\mathcal{H}$ and all x in Ω. Now define the bounded linear operator $Af : \mathcal{H} \to \mathbb{R}^n$, as

$$(Af)(x) = \langle f, K_x \rangle = f(x).$$

From which we obtain a discretized version $A_x f : \mathcal{H} \to \mathbb{R}^M$ of A in the following way

$$(A_x f)_i = \langle f, K_{a_i} \rangle = f(a_i)$$

where $\mathbb{R}^M$ has the inner product $(z, z')_{\mathbb{R}^M} = \sum_{i=1}^{M} z_i z_i'$. Then it is straightforward to see that

$$\|A_x f - z\| = \sum_{i=1}^{M} (f(a_i) - z_i)^2$$

Using this setting, we see surface reconstruction in terms of the formal definition (6.6) as the minimization of the functional $J_{\lambda, v_\delta}[f]$, given by

$$J_{\lambda, v_\delta}[f] = \|A_x f - z\|_{\mathbb{R}^M}^2 + \lambda R[f]$$

$$= \sum_{i=1}^{M} (f(a_i) - z_i)^2 + \lambda R[f]$$

One of the fundamental ideas of smoothing splines is that the mathematical theory of regularization provides a useful theory for incorporating prior knowledge, constraints, and quality of solution. This is commonly done by adding a suitable functional for each situation.

Chapter 7
3D Interpolation and Approximation

In this chapter we extend to n dimensions the spline theory developed in Chap. 4 on cubic splines for interpolation and approximation of functions; in particular we deal with 3D data. As a consequence of interpolation and smoothness conditions we obtain by a constructive proof, the Thin Plate Spline (TPS), whose explicit expression is given in terms of a convolution with the fundamental solutions of the biharmonic differential operator. In this construction no geometry on data is assumed. Thus, the data do not have to conform a regular grid, so we speak about reconstruction from scattered data.

7.1 Splines for 3D Data

TPS is a generalization to $\mathbb{R}^n$ of the well-known cubic spline in one variable [1, 3, 5, 12]. This spline was developed for solving interpolation problems in aircraft, ship building and automotive industries at the 1950's. Mathematicians soon realized that common interpolation methods, as Lagrange polynomials, were not suitable for tackling these problems. It was necessary to build more subtle tools. After this achievement, there was a great interest to obtain the $\mathbb{R}^n$ equivalent to cubic spline and several viewpoints were tried. The variational approach has shown to be the most versatile.

The problem we are going to solve now, is surface interpolation by thin plate spline. Given a cloud point $\mathcal{A} = \{a_1, a_2, ..., a_M\}$ sampled from a surface, each of this points have tree coordinates such that $a_i = (x_i, y_i, z_i)$. To reconstruct the surface, it is assumed that data comes from the sampling of a function $f : \mathbb{R}^2 \to \mathbb{R}$ such that f interpolates by $z_i = f(x_i, y_i)$ or approximates with $z_i \approx f(x_i, y_i)$. The goal is to approximate $f(\mathbf{x})$ when $\mathbf{x} \in \mathbb{R}^2$ does not belong to the set of nodes or centres $\mathcal{A}$.

Both cases, interpolation and approximation can be included in the regularization approach, where the regularization functional $R[f]$ is the deformation potential energy of a thin plate, given by

$$R[f] = \int_{\mathbb{R}^2} (f_{xx}^2 + 2f_{xy}^2 + f_{yy}^2) \, dy \, dx$$

H. Montegranario, J. Espinosa, *Variational Regularization of 3D Data*,
SpringerBriefs in Computer Science, DOI 10.1007/978-1-4939-0533-1_7,
© The Author(s) 2014

Thus the problem is to minimize the functional

$$J[f] = \sum_{i=1}^{M}(f(a_i) - z_i)^2 + \lambda R[f]$$

$$= \sum_{i=1}^{M}(f(\mathbf{x}) - z_i)^2 \delta\,(\mathbf{x} - a_i) + \lambda \int_{\mathbb{R}^2}(f_{xx}^2 + 2f_{xy}^2 + f_{yy}^2)d\mathbf{x}.$$

Calculating the first variation of $J[f]$ and simplifying

$$\sum_{i=1}^{M}(f(\mathbf{x}) - z_i)\delta(\mathbf{x} - a_i) + \lambda\Delta^2 f = 0,$$

$\Delta^2 f$ is the biharmonic operator. In this way is obtained the differential equation

$$\Delta^2 f = \sum_{i=1}^{M}\frac{(f(\mathbf{x}) - z_i)}{\lambda}\delta(\mathbf{x} - a_i)$$

with solution

$$f(x) = K * \sum_{i=1}^{M}\frac{(f(\mathbf{x}) - z_i)}{\lambda}\delta(\mathbf{x} - a_i),$$

applying the fundamental solutions of $\Delta^2 f$, we have

$$K(\mathbf{x}, \mathbf{y}) = \phi(\|\mathbf{x} - \mathbf{y}\|),\ \phi(r) = r^2\log r,$$

As in the case of cubic spline, the null space of $R[f]$ is $\Pi_1(\mathbb{R}^2)$, the set of linear polynomials in two variables of degree at most 1, we have to add a term $p(\mathbf{x}) = \beta_1 + \beta_2 x + \beta_3 y$ from this null space, to the former solution, obtaining the solution $S_\lambda(\mathbf{x})$ (or $S(\mathbf{x})$)

$$S_\lambda(\mathbf{x}) = \sum_{i=1}^{M}\alpha_i\phi(\|\mathbf{x} - a_i\|) + p(\mathbf{x}),\quad \phi(r) = r^2\log r \tag{7.1}$$

Although we can use any base for the polynomial terms, it is common to take the base $\mathbf{x}^\alpha = x_1^{\alpha_1}x_2^{\alpha_2}\cdots x_n^{\alpha_n}$, $\alpha = (\alpha_1, \alpha_2, ..., \alpha_n)$. Thus, for example, $\{1, x, y\}$, $\{1, x, y, x^2, y^2, xy\}$, $\{1, x, y, z, x^2, y^2, z^2, xy, xz, yz\}$ are basis for $\Pi_1(\mathbb{R}^2), \Pi_2(\mathbb{R}^2)$ and $\Pi_2(\mathbb{R}^3)$, respectively. In general, $\{\mathbf{x}^\alpha\}$, $\alpha \in Z^m$, $|\alpha| \le m$ is a basis for $\Pi_m(\mathbb{R}^n)$ with dimension

$$N = \binom{n+m}{n}.$$

Given the data $D = \{(a_i, z_i) \in \mathbb{R}^2 \times \mathbb{R}\}_{i=1}^M$, the interpolant S is completely determined by finding α_i's and β_i's. The interpolation conditions $S(a_i) = z_i$ $i = 1, 2, \ldots, M$, produce M equations and the remaining N degrees of freedom (in this case $N = 3$) are absorbed by the condition

$$\sum_{i=1}^N \alpha_i p_j(a_i) = 0 \quad \forall p_j \in \Pi_1(\mathbb{R}^2), \ j = 1, \ldots, N$$

obtaining the system

$$\begin{bmatrix} A + \lambda I & P \\ P^t & 0 \end{bmatrix} \begin{bmatrix} \alpha \\ \beta \end{bmatrix} = \begin{bmatrix} z \\ 0 \end{bmatrix} \tag{7.2}$$

where α_i, β_i are found solving this linear system, A is an interpolation matrix, with

$$A_{ij} = \phi(\| a_i - a_j \|) i, j = 1, \ldots, M,$$

$$P_{ij} = p_j(a_i), i = 1, \ldots, M, j = 1, \ldots, N,$$

$$\alpha = [\alpha_1, \alpha_2, \ldots, \alpha_M]^t,$$

$$z = [z_1, z_2, \ldots, z_M]^t,$$

Two conditions should hold for this system to have a unique solution for all values of the data D. Firstly, the interpolation matrix $A_{i,j} = \phi(\| a_i - a_j \|)$ should be non-singular over the subspace of vectors α satisfying $P^t \alpha = 0$.

Secondly, polynomials in $\Pi_m(\mathbb{R}^n)$ should be uniquely determined by their values on the set $\mathcal{A}$, that is, if $p \in \Pi_m(\mathbb{R}^n)$ and $p(a_i) = 0$, $i = 1, \ldots, M$ then $p = 0$. In this case $\mathcal{A}$ is said to be **unisolvent** with respect to $\Pi_m(\mathbb{R}^n)$.

Theorem 1 ([6]). If $m > \dfrac{n}{2}$, the D^m spline interpolation problem is well-posed: its solution exits, is unique, and depends continuously on the data $D = \{(a_i, z_i) \in \mathbb{R}^n \times \mathbb{R}\}_{i=1}^M$

7.2   Properties of Thin Plate Spline

$S_\lambda(\mathbf{x})$ is known as D^m spline, polyharmonic spline, Thin Plate Spline or surface spline. From a physical point of view the *regularization or penalty functional* $R[f]$ [12, 45] represents the bending energy of an infinitely extended plate. This spline interpolation, whether in one or two dimensions, physically corresponds to forcing a thin elastic beam or plate to pass through the data constraints. Away from the data points or centers the curve (or surface) will take on the shape that minimizes the strain energy given by $R[f]$

$S_\lambda(\mathbf{x})$ provides solution to interpolation and least squares approximation. It can be shown that $\lim_{\lambda \to \infty} S_\lambda(\mathbf{x})$ is the least squares regression onto the null space of $R[f]$ and $\lim_{\lambda \to 0} S_\lambda(\mathbf{x})$ is the interpolant to $f(a_i) = z_i$ that minimizes $R[f]$. With too little regularization (λ too small), the reconstruction have highly oscillatory artifacts due to noise amplification. With too much regularization (λ very large), the reconstruction is too smooth. So it is very relevant to find methods for the optimal value of λ. Other constraints can be introduced in the optimization scheme, for example in some problems f should be positive; the least squares term can be replaced by other measures of fit to data or other penalty functionals $R[f]$ can be introduced. However, what makes $S_\lambda(\mathbf{x})$ most useful for applications is the fact that the interpolation problem becomes insensitive to the dimension n of the space $\mathbb{R}^n$ in which the data sites lie, thanks to the radial representation as translates of a single basis function. Instead of having to deal with a multivariate function (whose complexity will increase with increasing space dimension n) we can work with the same univariate function $\phi(r)$ for all choices of n. This framework can be generalized to $\mathbb{R}^n$, taking as regularization functional the seminorm

$$R[f] = \int_{\mathbb{R}^n} \sum_{|\alpha|=m} \frac{m!}{\alpha!} |\partial^\alpha f(\mathbf{x})|^2 \, d\mathbf{x},$$

which null space is $\Pi_m(\mathbb{R}^n)$. Applying regularization, the result is the same spline as $S_\lambda(\mathbf{x})$ and $\Phi(r)$ is chosen from the fundamental solutions of the iterated Laplacian operator $\Delta^m f$

$$\phi(r) = \begin{cases} cr^{2m-n} \ln r, & n \text{ even} \\ dr^{2m-n}, & n \text{ odd} \end{cases}$$

Tables 7.1 and 7.2 show some of the possible cases for the penalty functional and the polynomial term

Example 1 As a first test for TPS interpolation, we interpolate a set of points on Franke's function (Fig. 7.1), using scattered data $D = \{(a_i, z_i) \in \mathbb{R}^2 \times \mathbb{R}\}_{i=1}^M$. for different values of M. The linear system to solve is

Table 7.1 Thin plate seminorm for some values of m, n ($m = 2,3; n = 1,2,3$) m represents the degree of derivatives and n the dimension of data

$\mathbb{R}^n$	D^m	$\|u\|^2_{D_m} = \int_{\mathbb{R}^n} \sum_{|\alpha|=m} \dfrac{m!}{\alpha!} |\partial^\alpha u|^2 \, dx$	$u : \mathbb{R}^n \to \mathbb{R}$
$\mathbb{R}^1$	D^2	$\int (u''(x))^2 \, dx$	
$\mathbb{R}^1$	D^3	$\int (u'''(x))^2 \, dx$	
$\mathbb{R}^2$	D^2	$\iint_{\mathbb{R}^2} \left[\left(\dfrac{\partial^2 u}{\partial x^2}\right)^2 + 2\left(\dfrac{\partial^2 u}{\partial x \partial y}\right)^2 + \left(\dfrac{\partial^2 u}{\partial y^2}\right)^2 \right] dx\, dy$	
$\mathbb{R}^2$	D^3	$\iint_{\mathbb{R}^2} \left[\left(\dfrac{\partial^3 u}{\partial x^3}\right)^2 + 3\left(\dfrac{\partial^3 u}{\partial x^2 \partial y}\right)^2 + 3\left(\dfrac{\partial^3 u}{\partial x \partial y^2}\right)^2 + \left(\dfrac{\partial^3 u}{\partial y^3}\right)^2 \right] dx\, dy$	
$\mathbb{R}^3$	D^2	$\iint_{\mathbb{R}^2} \left[\left(\dfrac{\partial^2 u}{\partial x^2}\right)^2 + \left(\dfrac{\partial^2 u}{\partial y^2}\right)^2 + \left(\dfrac{\partial^2 u}{\partial z^2}\right)^2 + 2\left(\dfrac{\partial^2 u}{\partial x \partial y}\right)^2 + 2\left(\dfrac{\partial^2 u}{\partial x \partial z^2}\right)^2 + 2\left(\dfrac{\partial^2 u}{\partial y \partial z}\right)^2 \right] dx\, dy\, dz$	

Table 7.2 Polynomial terms for D^m splines

$\mathbb{R}^n$	m	$p(\mathbf{x})$
$\mathbb{R}^1$	2	$\beta_0 + \beta_1 x$
$\mathbb{R}^1$	3	$\beta_0 + \beta_1 x + \beta_2 x^2$
$\mathbb{R}^2$	2	$\beta_0 + \beta_1 x + \beta_2 y$
$\mathbb{R}^2$	3	$\beta_0 + \beta_1 x + \beta_2 y + \beta_3 x^2 + \beta_4 y^2 + \beta_5 xy$
$\mathbb{R}^3$	2	$\beta_0 + \beta_1 x + \beta_2 y + \beta_3 z$

$$
\begin{bmatrix}
A_{11} & A_{12} & \cdots & A_{1M} & 1 & x_1 & y_1 \\
A_{21} & A_{22} & \cdots & A_{2M} & 1 & x_2 & y_2 \\
\vdots & \vdots & \ddots & \vdots & \vdots & \vdots & \vdots \\
A_{M1} & A_{M2} & \cdots & A_{MM} & 1 & x_M & y_M \\
1 & 1 & \cdots & 1 & 0 & 0 & 0 \\
x_1 & x_2 & \cdots & x_M & 0 & 0 & 0 \\
y_1 & y_2 & \cdots & y_M & 0 & 0 & 0
\end{bmatrix}
\begin{bmatrix}
\alpha_1 \\ \alpha_2 \\ \vdots \\ \alpha_M \\ \beta_1 \\ \beta_2 \\ \beta_3
\end{bmatrix}
=
\begin{bmatrix}
z_1 \\ z_2 \\ \vdots \\ z_M \\ 0 \\ 0 \\ 0
\end{bmatrix},
\tag{7.3}
$$

where the coordinates of observation points are $a_i = (x_i, y_i)$, $A_{ij} = \phi(\| a_i - a_j \|)$. We solve this system to get the coefficients or weights α_i, β_j $i = 1,\ldots,M$, $j = 1,2,3$.

Example 2 The following is the mathematical expression for Franke's data surface in the interval $\Omega = [0,1] \times [0,1]$ The results of the interpolation are shown in (Fig. 7.2).

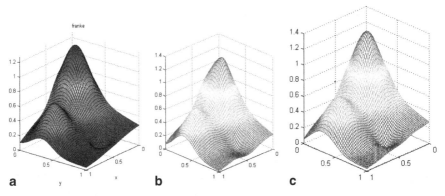

Fig. 7.1 Results interpolating Franke's test function by thin plate spline without noise in the interval $\Omega = [0,1] \times [0,1]$ **a** Original function. Interpolation with: **b** $M = 40$ points. **c** $M = 100$ points. The interpolation rapidly improves with an increasing number of random points

$$
f(x,y) = \begin{cases}
\dfrac{1}{2} & y \le \dfrac{2}{5} \\[2mm]
\dfrac{1}{2}\left(1 - \dfrac{25}{9}(y - \dfrac{2}{5})^2\right) & y > \dfrac{2}{5} \wedge x \le \dfrac{1}{5} \\[2mm]
\dfrac{125}{72}(1-y)^2(1-x) & y > \dfrac{2}{5} \wedge x > \dfrac{1}{5}
\end{cases}
$$

A fundamental property of $S(\mathbf{x})$ is that as the set $\mathcal{A}$ fills Ω (increasing M) the error between the function and its interpolant should go to zero. The usual measure for the way $\mathcal{A}$ fills out Ω, is $h = \sup_{\mathbf{x} \in \Omega} \inf_{a \in \mathcal{A}} \| \mathbf{x} - a \|$. If f is a continuous function then $\| S - f \| = O(h^r)$ as $h \to 0$, where r is a measure of the smoothness of f [72]. The thin plate spline $S_\lambda(\mathbf{x})$ has the important property of being a radial basis function.

Definition 1 A function $\Phi : \mathbb{R}^n \to \mathbb{R}$ is called radial if there exists a univariate function $\phi : [0, \infty) \to \mathbb{R}$ such that

$$
\Phi(\mathbf{x}) = \phi(r), \text{ with } r = \| \mathbf{x} \|,
$$

where $\|\cdot\|$ is the Euclidian norm and ϕ is called a **radial basis function(RBF)**.

Usually RBF's can be constructed for scattered data $D = \{(a_j, z_j)\}_{j=1}^M$ by defining a kernel function K

$$
K : \Omega \times \Omega \to \mathbb{R}
$$

which in the case of multivariate function approximation depends on $\mathcal{A} = \{a_1, a_2, \dots, a_M\} \subset \Omega \subset \mathbb{R}^n$. K is translational and rotational invariant, by using a radial function Φ in the form

$$
K(\mathbf{x}, \mathbf{y}) = \Phi(\mathbf{x} - \mathbf{y}) = \phi(\| \mathbf{x} - \mathbf{y} \|)
$$

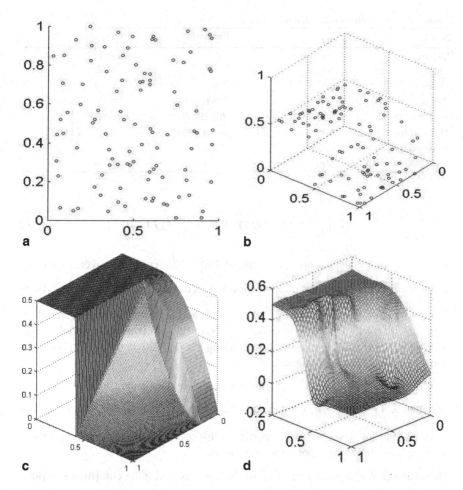

Fig. 7.2 **a** (x, y) coordinates of 100 scattered data on Franke's data surface, **b** 3D scatter plot of the points, **c** The original surface, **d** Interpolation of the scattered data. This is an extreme case very useful for testing approximation models in the reproduction of discontinuities. It has smooth zones as well as vertices, edges and faces

A classic example is the Gaussian kernel, $K(\mathbf{x}, \mathbf{y}) = \exp(\| \mathbf{x} - \mathbf{y} \|_2^2)$. Table 7.3 gives a list of the most common radial basis functions.

7.3 Multivariate Approximation and Positive Definiteness

In the theory of integral equations it is common to find operators in the form $(\mathcal{K} f)(\mathbf{x}) = \int_\Omega K(\mathbf{x}, \mathbf{y}) f(\mathbf{y}) d\mathbf{y}$. A complex valued continuous kernel $K(\mathbf{x}, \mathbf{y})$ is positive definite when $K(\mathbf{x}, \mathbf{y})$ satisfies

Table 7.3 Some well known radial basis functions

Name	$\phi(r)$	Parameters	Order
Linear	$\phi(r) = r$		$m = 1$
Cubic	$\phi(r) = r^3$		$m = 2$
Gaussian	$\phi(r) = e^{-\beta r^2}$	$\beta > 0$	$m = 0$
Poli-harmonic	$\phi(r) = r^\beta$	$\beta > 0, \beta \notin 2\mathbb{N}$	$m \geq \lceil \beta/2 \rceil$
Thin plate spline	$\phi(r) = r^\beta \log(r)$	$\beta \in 2\mathbb{N}, \beta > 0$	$m \geq \lceil \beta/2 \rceil$
Multiquadric	$\phi(r) = (c^2 + r^2)^{\beta/2}$	$\beta > 0 \quad \beta \notin 2\mathbb{N}$	$m \geq \lceil \beta/2 \rceil$
Inverse multiquadric	$\phi(r) = (1 + r^2)^{\beta/2}$	$\beta < 0$	$m = 0$
Wendland function	$\phi(r) = (1-r)_+^4 (4r+1)$	$n \leq 3$	$m = 0$

$$\int_\Omega \int_\Omega K(\mathbf{x}, \mathbf{y}) f(\mathbf{x}) \overline{f(\mathbf{y})} d\mathbf{x} d\mathbf{y} = (\mathcal{K} f, f) \geq 0, \tag{7.4}$$

for any continuous function $f(\mathbf{x})$. Around 1909, Mercer [2] discovered that this condition is equivalent to the quadratic form $\sum_{i,j=1}^{M} K(a_i, a_j) \xi_i \overline{\xi_j}$ to be positive definite (≥ 0) for any M points $a_1, a_2, \cdots, a_M \in \mathbb{R}^n$ and arbitrary complex numbers $\xi_1, \xi_2, \ldots, \xi_M$. It can also be shown that if $K(\mathbf{x}, \mathbf{y})$ in (7.4) is positive definite, then $K(\mathbf{x}, \mathbf{y})$ is Hermitian symmetric. Later during the 1920's, J. Moore [2] used the name "positive Hermitian matrices."

Given any function $\Phi : \mathbb{R}^n \to \mathbb{C}$, we can associate with it some kind of interpolation matrix $A_{jk} = \Phi(a_j - a_k)$ and the quadratic form $\sum_{j=1}^{M} \sum_{k=1}^{M} \alpha_j \overline{\alpha}_k \Phi(a_j - a_k)$. The following definition describes the properties of this quadratic form and its relationship with surface splines and approximation theory.

Definition 2 A continuous function $\Phi : \mathbb{R}^n \to \mathbb{C}$ is said to be **conditionally positive definite** (CPD) of order m on $\mathbb{R}^n$ if

$$\boldsymbol{\alpha}^t A \boldsymbol{\alpha} = \sum_{j=1}^{M} \sum_{k=1}^{M} \alpha_j \overline{\alpha}_k \Phi(a_j - a_k) \geq 0$$

for any M different points $\mathcal{A} = \{a_1, a_2, \ldots, a_M\} \subseteq \Omega \subset \mathbb{R}^n$ and $\boldsymbol{\alpha} = [\alpha_1, \alpha_2, \ldots, \alpha_M]^t \in \mathbb{C}^n$, satisfying

$$\sum_{j=1}^{M} \alpha_j p(a_j) = 0, \forall p \in \Pi_{m-1}(\mathbb{R}^n),$$

If $\boldsymbol{\alpha}^t A \boldsymbol{\alpha} > 0$ whenever the points a_i are distinct and $\boldsymbol{\alpha} \neq 0$, then we say that the function Φ is **strictly conditionally positive definite** of order m. In particular, the case $m = 0$ yields the class of (strictly) positive definite functions. As a consequence of this definition a function which is CPD of order m on $\mathbb{R}^n$ is also CPD of any higher order. Micchelli [48] showed that interpolation with strictly CPD functions of order one is possible even without adding a polynomial term. The importance of these functions in approximation theory is illustrated in the following

Theorem 2 Let $\Phi : \mathbb{R}^n \to \mathbb{C}$ be conditionally positive definite of order m on $\Omega \subset \mathbb{R}^n$ and let the data $\mathcal{A} = \{a_1, a_2, \ldots, a_M\} \subseteq \Omega$ be Π_m-nondegenerate (i.e zero is the only polynomial that vanishes on $\mathcal{A}$). Then the system (7.2) is uniquely solvable [72].

Chapter 8
Radial Basis Functions

This last chapter may be seen as a general introduction to meshless methods not only for surface reconstruction but in other problems of scattered data. The general interpolation and smoothing problems are described in terms of radial basis functions; we then illustrate some characterizations of these functions. The theory is applied to noisy data by tuning the regularization parameter with generalized cross validation and we finally give some examples illustrating the theory.

8.1 Meshless Methods

From the point of view of practical implementations the methods in this book can be seen as Radial basis functions applications and they are part of what is known today as meshless methods. They first appeared in the context of applications in geosciences such as meteorology, geo-physics and mapping; disciplines in which is necessary to make interpolations over scattered data. These methods can be applied to data defined on irregular grids, an advantage over standard multivariate approximation methods (e.g. finite elements) that require an underlying mesh or triangulation for definition of basis functions or elements.

Meshless methods appear during the second half of twentieth century. From this time, late 1960s, comes the method of Shepard functions [12] for surface modelling and the study of Hardy's Multiquadrics [30]. Franke [17, 18] conjectured the invertibility of the interpolation matrix for the multiquadric, a fact that was proved later by Micchelli [48].

Rotation invariant seminorms [12] were studied by Duchon, who applied an abstract variational approach on distribution spaces, obtaining convolution expressions for the now well-known thin plate spline. Using the Sobolev embedding theorem, he found that his spaces are included into the space of continuous functions $C(\mathbb{R}^n)$, making possible the work of interpolation.

During the last decades these methods have spread into many disciplines such as, artificial intelligence, computer vision, neural networks, sampling theory, Geostatistics (kriging, cokriging), numerical solution of PDE's and learning theory.

H. Montegranario, J. Espinosa, *Variational Regularization of 3D Data*, 69
SpringerBriefs in Computer Science, DOI 10.1007/978-1-4939-0533-1_8,

RBF's can be classified by their order m. If $m = 0$ the function is positive definite, for example the Gaussian. Conditionally positive definite functions have order $m \geq 1$. When $A = \{a_1, a_2, \ldots, a_M\}$ is unisolvent with respect to polynomials on $\mathbb{R}^n$ with total degree less than m ($\Pi_m(\mathbb{R}^n)$), then, for data $f|_A$, generated by a continuous function f, there is a unique interpolant

$$S(x) = \sum_{a_i \in A} \alpha_i \phi(\| x - a_i \|) + p(x) \tag{8.1}$$

that satisfy $f|_A = S|_A$. If f is a polynomial in $\Pi_m(\mathbb{R}^n)$, then $S = f$. That is, the interpolant can reproduce polynomials. The expression for $S(x)$ permits to work efficiently for large dimensions n, because the function ϕ reduces the problem to one dimensional evaluations. In general, we can use radial basis functions for solving two problems:

P1. *Interpolation problem*

$$\min_{u \in \mathcal{U}} J[u], \quad \text{with } \{u(a_i) = z_i\}_{i=1}^M$$

P2. *Smoothing problem*

$$\min_{u \in \mathcal{U}} R_\lambda[u] = \sum_{a_i \in A} (u(a_i) - z_i)^2 + \lambda J[u], \quad \text{with } \{u(a_i) \approx z_i\}_{i=1}^M$$

and both problems are solved by the linear system

$$\begin{bmatrix} A + \lambda I & P \\ P^t & 0 \end{bmatrix} \begin{bmatrix} \alpha \\ \beta \end{bmatrix} = \begin{bmatrix} z \\ 0 \end{bmatrix}, \tag{8.2}$$

with $\lambda = 0$ for the interpolation problem.

8.2   Radial Basis Functions and Scattered Data

The crucial question to be answered when we want to apply an approximation model is about the invertibility of the interpolation matrix A. Commonly, the approximator $S(x)$ is a linear combination of certain basis functions $u_k(x)$

$$S(x) = \sum_{k=1}^N \alpha_k u_k(x), \quad x \in \mathbb{R}^n$$

The application of interpolation conditions yields a linear system of the form $A\alpha = z$; then this problem is well-posed if and only if the matrix A is invertible. In the one dimensional case it is possible to interpolate arbitrary data by means of polynomials.

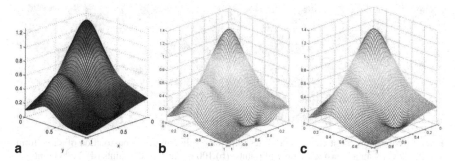

Fig. 8.1 (a) Franke's function and reconstruction with (b) Gaussian, $\beta = 25$ and (c) Inverse multiquadric $\varepsilon = 5$

Nevertheless, this is not so easy in further dimensions. The following theorem provides an answer.

Theorem 1. (Mairhuber-Curtis) Let $\mathcal{U}$ be an M-dimensional space of continuous real-valued functions and some domain $\Omega \subset \mathbb{R}^n$ and assume that any set $A = \{a_1, a_2, \ldots, a_M\} \subseteq \Omega$ is $\mathcal{U}$-nondegenerate. Then either $M = 1$ or $n = 1$ hold. That is, either the function space or the underlying domain are just one dimensional.

As a consequence, it is not possible to obtain a unique interpolation with bivariate polynomials of degree M on data given at arbitrary locations in $\mathbb{R}^2$ and the situation is similar for higher dimensions. This also says that the multivariate scattered data interpolation problem is well-posed if the basis functions depend on the data locations $A = \{a_1, a_2, \ldots, a_M\}$.

RBF's have been widely studied on their performance and properties. One of the first results about this goes back to Franke's [18], who analysed a large number of multivariable interpolation methods, concluding that the best functions, from the analysed set, were the multiquadric and thin plate spline.

Radial functions may have global or compact support. Due to sparseness of the corresponding interpolation matrices, compactly supported functions [6, 55] offer computational advantages. If we need functions with strictly local effects, we should use small supports, but the global behavior cannot be properly recovered, so a possible strategy is to combine both type of functions.

Example 1 This example shows the performance of some radial basis functions (Fig. 8.1) for interpolation, taking $M = 100$ random points on the Franke's surface as result of evaluating $z = f(x, y)$ in the region $\Omega = [0,1] \times [0,1]$.

$$f(x, y) = 0.75e^{-\frac{1}{4}((9x-2)^2 + (9y-2)^2)} + 0.75e^{-\frac{(9x+1)^2}{49} + \frac{(9y+1)}{10})}$$

$$+ 0.5e^{-(9x-7)^2 - \frac{(9y-3)^2}{4}} + 0.2e^{-((9x-4)^2 + (9y-7)^2)}.$$

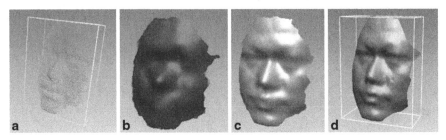

Fig. 8.2 We use the point cloud in (**a**) to test thin plate spline in the reconstruction of a free form object, with and an increasing number of points: (**b**) 100 points (**c**) 800 points (**d**) 2,000 points

In this case we have taken $\lambda = 0$ and the basis evaluated are (i) inverse multiquadric $\phi(r) = 1 / \sqrt{1 + \varepsilon^2 r^2}$ (ii) thin plate spline $\phi(r) = r^2 \log(r)$ and (iii) Gaussian $\phi(r) = \exp(-\beta r^2)$. $S(x)$ does not have polynomial term $p(x)$ for Gaussian and multiquadric, in which case the system to solve is $A\alpha = z$.

Example 2 Illustration of the convergence of thin plate spline reconstruction of a free form object using an increasing number of points (Fig. 8.2).

8.3   Characterizations of Radial Basis Functions

From the practical point of view the properties of RBF's can be summarized in the following

- They may be generalized to many kinds of data and applications
- They may be applied to scattered data on any dimension of $\mathbb{R}^n$
- It is possible to obtain for them explicit mathematical expressions as linear combinations of translates of a basis function whose weights are found solving a system of linear equations
- It is not difficult to find error bounds using seminorms and inner products on Hilbert spaces

These evident advantages and the initial success of radial basis functions like multiquadrics and thin plate spline, have encourage the community of approximation theory [16] to systematize their properties and include them into a general framework. Next we mention some of these characterizations, based on the concept of (conditionally) positive definite function.

8.3.1   Bochner's Characterization

This approach provides translation invariant positive definite functions in $\mathbb{R}^n$ thanks to a famous result of Bochner that establishes a close relationship between Fourier transforms and positive definite functions

$$\hat{f}(\xi) = \int_{-\infty}^{\infty} f(t)e^{i\xi t} dt$$

$f(x) = 1/\pi(1+x^2)$	$\hat{f}(x) = e^{-|x|}$
$f(x) = e^{-|x|}/2$	$\hat{f}(x) = 1/(1+x^2)$
$f(x) = \pi^{-1/2}e^{-x^2}$	$\hat{f}(x) = e^{-x^2/4}$
$f(x) = (1+x^{-2})/2\pi$	$\hat{f}(x) = |x|^{-1}(1-e^{-|x|})$
$f(x) = \begin{cases} 1/2 & |x| \le 1 \\ 0 & |x| > 1 \end{cases}$	$\hat{f}(x) = \sin x/x$

Table 8.1 A Fourier transform table. An easy way to find a great number of positive definite functions. (Taken from [7])

Theorem 2 (Bochner) A continuous function $\Phi : \mathbb{R}^n \to \mathbb{C}$ is a positive semi definite function if and only if it is the Fourier transform of a finite non-negative Borel measure on $\mathbb{R}^n$.

This theorem gives us, unexpectedly, an abundance of possible radial basis functions that produce invertible interpolation matrices. For obtaining some of them it is enough to consider a Fourier transform Table (Table 8.1).

In spite of the great number of available functions to be applied as radial basis, there is a relatively small number that have arose in particular problems and constitutes the most commonly used radial functions.

8.3.2 Schoenberg-Micchelli Characterization

This section derives the properties of functions conditionally positive definite function of order m, $\Phi(x,y) = \phi(\|x-y\|)$ on $\mathbb{R}^n \times \mathbb{R}^n$ from the properties of univariate functions $\phi : [0, \infty[\to \mathbb{R}$. In this case we say that ϕ is also CPD of order m

Definition A function $\phi : [0, \infty[\to \mathbb{R}$ is said to be completely monotone on] if $\phi \in C^\infty(0, \infty)$ and $(-1)^k \phi^{(k)}(r) \ge 0\ k \in \mathbb{N}\ r > 0$

A function $\phi : [0, \infty[\to \mathbb{R}$ is said to be completely monotone on $[0, \infty[$ if it is completely monotone on $]0, \infty[$ and continuous at zero.

Although with different intentions, one of the first results applicable in the interpolation problem was obtained by Schoenberg [58]:

Theorem 3. (Schoenberg) Suppose $\phi : [0, \infty[\to \mathbb{R}$ is not the constant function. Then ϕ is positive definite on every $\mathbb{R}^n$ if and only if the function $\phi(\sqrt{t}) t \in [0, \infty[$ is completely monotone on $[0, \infty[$

As a consequence, the interpolation matrix A is positive definite and in particular is non-singular for all distinct points $A = \{a_1, a_2, \dots a_M\}$ in $\mathbb{R}^n$. Schoenberg's characterization of positive definite functions allows to prove positive definiteness of Gaussians and inverse multiquadrics without difficulty. For example for the Gaussian $\phi(r) = e^{-\beta r^2}$, $\beta > 0$, we have

$$f(r) = \phi(\sqrt{t}) = e^{-\beta r},$$

which satisfies $(-1)^k f^{(k)}(r) = \beta^k e^{-\beta r} > 0$ for all $k \in \mathbb{N}$ and $\beta, r > 0$, then by the former theorem the Gaussian is positive definite. The Gaussian is the most remarkable example of completely monotonic function, further in some sense all completely monotonic functions are generated by integration of a measure with the Gaussian as a kernel. A theorem of Berstein [72], shows that a function ϕ is positive definite on $\mathbb{R}^n$ for all $n \geq 1$, if and only if there exists a nonzero, finite, nonnegative Borel measure μ, not supported in zero, such that ϕ is of the form

$$\phi(r) = \int_0^\infty e^{-r^2 t} d\mu(t)$$

This result implies that the proper tool to handle positive definite functions on $\mathbb{R}^n$ for all $n \geq 1$, is the Laplace Transform

8.3.3 Variational Characterization

Madych and Nelson generalized the ideas of Duchon, establishing a close relationship between conditionally positive definite functions and variational theory, applicable not only to thin plate spline but also to nonhomogeneous functions like multiquadrics. This approach reinterprets the interpolants as the minimization of a seminorm. It begins selecting an integer $m \geq 0$ and a continuous function $\Phi : \mathbb{R}^n \to \mathbb{C}$ conditionally positive definite of order m. Using Φ is constructed a space $\mathcal{X}$ with a semi-inner product $[\cdot, \cdot] : \mathcal{X} \times \mathcal{X} \to \mathbb{R}$ which satisfies all the properties of an inner product, but its null space may be different from zero. By construction $\mathcal{X}$ is a subspace of $C(\mathbb{R}^n)$ and the null space of $[\cdot, \cdot]$ is $\Pi_{m-1}(\mathbb{R}^n)$, the set of polynomials in $\mathbb{R}^n$ of total degree $\leq m-1$.

In this conditions is obtained an interpolating $S(x)$ in the form (8.1) such that S minimize the seminorm defined by $[.,.]$. For example, in the case of thin plate spline we may consider the seminorm

$$J[u] = |u|_m^2 = \int_{\mathbb{R}^n} \sum_{|\alpha| = m} \frac{m!}{\alpha!} |\partial^\alpha u(x)|^2 dx$$

With null space $\Pi_{m-1}(\mathbb{R}^n)$ of dimension N, as coming from the semi-inner product

$$[u, v] = \int_{\mathbb{R}^n} \sum_{|\alpha| = m} \frac{m!}{\alpha!} \partial^\alpha u(x) \partial^\alpha v(x) dx$$

To reconstruct a function f, suppose $\mathcal{A}$ contains a $\Pi_{m-1}(\mathbb{R}^n)$- unisolvent subset $\mathcal{B} = \{a_j\}_{j=1}^N$, and we want to find $S(\mathbf{x})$ such that $S(a_i) = f(a_i) = z_i$ for $i = 1,\ldots,M$ (or $S|_A = f|_A$) and minimize the spline seminorm.

The method begins assuming $\mathcal{X}$ vector space of all distributions for which all the partial derivatives of total order m are square integrable in $\mathbb{R}^n$ then by Sobolev embedding theorem [59], $\mathcal{X}$ is a linear subspace of $C(\mathbb{R}^n)$ for $m > \dfrac{n}{2}$. The key idea is to modify the semi inner product $[\cdot,\cdot]$ on $\mathcal{X}$. We then obtain an inner product for building a complete space $\mathcal{H}$. Once this is done, we are enabled to use all the machinery of Hilbert spaces. The inner product is defined as

$$(u,v) = [u,v] + \sum_{a_i \in B} u(a_i)v(a_i) \tag{8.3}$$

The solution $S(\mathbf{x})$ to the minimal norm interpolant is found using the projection operator technique on Hilbert Spaces with this new inner product in the following scheme. Let $p_1, p_2,\ldots, p_N$ be the Lagrange basis for $\Pi_{m-1}(\mathbb{R}^n)$ with respect to the points $\mathcal{B} = \{a_1, a_2,\ldots, a_N\}$, define the projection $P: \mathcal{H} \to \mathcal{H}$ by $Pf = \sum_{i=1}^N f(a_i)p_i$. Then $\Pi_{m-1}(\mathbb{R}^n)$ is the range of P and $\mathcal{H}_0 = \{u \in \mathcal{H}: u(a_j) = 0, \forall a_j \in \mathcal{B}\}$ its null space, then $\mathcal{H}_0 \perp \Pi_{m-1}(\mathbb{R}^n)$ and $H = H_0 \oplus \Pi_{m-1}(\mathbb{R}^n)$, using the Riesz representation theorem it is possible to arrive at the results obtained in former chapters by a very different way [37, 38]

8.4 Regularization and Noise

Noise is intrinsic and ubiquitous in measured data. It is a disturbance that affects a signal and may have many sources. Measurement noise is often assumed to be Gaussian in a wide range of disciplines. Radial basis functions can be applied to exact or noisy data. Nevertheless, in this second case, are necessary some additional tasks for improving the approximation. It is a well known fact in the literature that the same functions we have used in the examples above may suffer a surprising failure when applied on noisy data without choosing appropriate values for parameters.

Given a set of noisy scattered data on a surface, there will be many possible surfaces consistent with these initial points. The resulting function f that models the surface should be such that it generalizes well, that is, it should give useful values of $f(\mathbf{x})$ to new points $\mathbf{x} \notin \mathcal{A}$. Furthermore, it should be *stable* in the sense that small changes in the training data do not change f too much. That is, we need a careful balance between generalization and stability properties on one hand and data reproducing quality on the other. This is called the *reproduction-generalization dilemma*.

An important task for dealing with these problems is the determination of the smoothing parameter λ. This is a very sensitive parameter, small changes in its values may cause large changes in the reconstruction of the surface. λ is penalizing large values of the functional $R[f]$, which contains a large number of derivatives such that should assume large values for non smooth functions and small for smooth functions. It is possible to find optimal values $\hat{\lambda}$ for the smoothing parameter in the sense of some criteria. One of them is **cross validation**.

The basic idea of cross validation for evaluating the performance of an approximator of the data $D = \{(a_i, z_i) \in \mathbb{R}^n \times \mathbb{R}\}_{i=1}^{M}$ is to build the approximator over the set $\{a_i : i \neq k\}$ that is, excluding or leaving out the knot a_k and then repeating the procedure for $k = 1, \ldots, M$. Let $S_\lambda^{[k]}$ be the minimizer of

$$J[u] = \sum_{\substack{i=1 \\ i \neq k}}^{M} (u(a_i) - z_i)^2 + \lambda R[u],$$

then it is reasonable to look for the model which minimizes the ordinary cross validation function $V_0(\lambda)$, defined as

$$V_0(\lambda) = \frac{1}{M} \sum_{k=1}^{M} (z_k - S_\lambda^{[k]}(a_k))^2,$$

The minimizer of $V_0(\lambda)$ is known as the "leaving-out-one" estimator of λ but in computational terms, finding this value is very expensive. However there exist an even easier method called Generalized Cross Validation (GCV) [69, 70], that uses the SVD with diagonal elements $d_1 \geq d_2 \geq \cdots \geq d_p \geq 0$, to obtain the expression

$$V(\lambda) = M \sum_{j=1}^{M-t} \left(\frac{n\lambda}{d_j^2 + n\lambda} \right)^2 Z_j^2 \Bigg/ \left(\sum_{j=1}^{M-t} \frac{n\lambda}{d_j^2 + n\lambda} \right)^2 \qquad (8.4)$$

Example 3. Interpolation and Cross validation The results for this example are shown in Fig. 8.3. For this experiment we have taken 400 noisy points $\mathcal{A} = \{a_1, a_2, \ldots, a_M\}$ on Franke's function in $\Omega = [0,1] \times [0,1]$ to be approximated with Thin Plate Spline. We can allow the surface to pass close to, but not necessarily through, the known data points, by setting $\lambda > 0$. When $\lambda = 0$, the function interpolates the data points. As λ approaches zero, the surface becomes rougher because it is constrained to pass closer to the data points. At $\lambda = 0$, the surface interpolates the data, and overshoots are much more evident. The optimum value λ_0 for λ is determined minimizing the GCV function $V(\lambda)$ to obtain a smooth surface with fidelity to data. At larger values of λ ($\lambda > \lambda_0$), the reconstructed model is smoother and approaches an amorphous bubble.

8.5   How to Judge an Approximation

There are three very useful criteria to judge effectiveness of an approximator: *density, interpolation* and *order of convergence*. RBF's fit all these requirements.

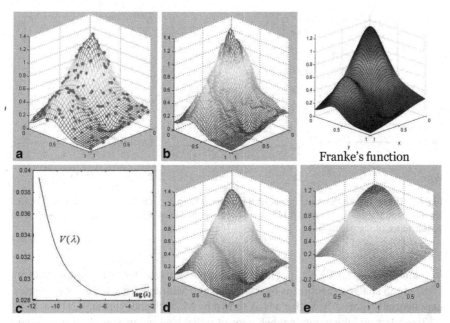

Fig. 8.3 This example shows the potential of splines to smooth a noisy surface by tuning regularization parameters with Generalized Cross- validation (GCV). (**a**) Depicts the noisy data near the surface. In (**b**) we see the interpolation of the original data with $\lambda = 0$, without smoothing. (**c**) The optimum value λ_0 for the regularization parameter λ is found by minimizing the function $V(\lambda)$ and the resulting surface is shown in (**d**). When the surface is reconstructed by using values larger than λ_0 the surface is too smoothed and loses fidelity to data as it is the case shown in (**e**)

A subset X of a topological space $\mathcal{X}$ called **dense** (in $\mathcal{X}$) if every point x in $\mathcal{X}$ either belongs to X or is a limit point of X.

A subset U in a normed space $\mathcal{U}$ is said to be **fundamental** if the set of all linear combinations from U is dense in $\mathcal{U}$. Otherwise, $\forall f \in \mathcal{U}$ and $\forall \varepsilon > 0$ there is a vector $\sum_{k=1}^{N} \alpha_k u_k$ with $u_k \in U$, such that

$$\left\| f - \sum_{k=1}^{N} \alpha_k u_k \right\| \leq \varepsilon.$$

Error bounds refers to the size, conditions and properties of ε in this inequality.

The convergence of interpolants on scattered data can be studied in terms of the spatial density of the set of nodes $\mathcal{A} = \{a_1, a_2, \ldots a_M\}$, by the Hausdorff or fill distance

$$h_{A,\Omega} = \sup_{x \in \Omega} \min_{a_j \in A} \| x - a_j \| \tag{8.5}$$

of the set $\mathcal{A}$ within an enclosing domain Ω. $h_{A,\Omega}$ gives the radius of the largest ball without data sites or "data-site free ball" in Ω.

As the set $\mathcal{A}$ "fills" Ω, the error between the function and its interpolant goes to zero. If $h_{A,\Omega}$ tends to zero, the reproduction error always behaves like a power

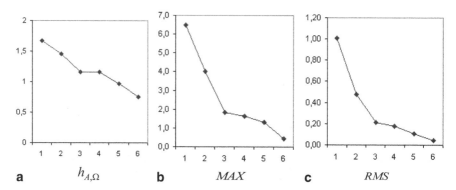

Fig. 8.4 Illustration of the convergence of Radial basis functions. (**b**) The Maximum Error MAX and (**c**) Root Mean Square Error RMS decrease when the Hausdorff measure (**a**) $h_{A,\Omega} \rightarrow 0$ as a consequence of increasing M

$h_{A,\Omega}^k$ and $\| f - S_f \| = \mathcal{O}(h_{A,\Omega}^k)$, where $k > 0$ increases with the smoothness of $\phi(r)$. If $\phi(r)$ is an analytic function as the Gaussian and Multiquadrics, the error decreases exponentially like $ce^{-c/h}$ with $c > 0$ as shown by Madych and Nelson [42] (Fig. 8.4).

To understand the numerical behavior of radial basis approximations it is essential to have bounds on the approximation error and on the condition numbers of the interpolation matrix in the system

Associated with the solution of a linear system $A\mathbf{x} = \mathbf{b}$ it is common to use the condition number $\kappa(A)$ of the matrix A defined as a product of the magnitudes of A and its inverse; that is

$$\kappa(A) = \| A \| \cdot \| A^{-1} \|, \tag{8.6}$$

where $\| \cdot \|$ is a matrix norm. If the solution of $A\mathbf{x} = \mathbf{b}$ is insensitive to small changes in the right-hand side $\mathbf{b}$ then small perturbations in b result in only small perturbations in the computed solution $\mathbf{x}$. In this case, A is said to be well conditioned and corresponds to small values of $\kappa(A)$. If the condition number is large then A is ill conditioned and the numerical solutions of $A\mathbf{x} = \mathbf{b}$ cannot be trusted.

The price we have to pay for the great properties of radial basis functions is that their interpolation matrices may become strongly ill conditioned, resulting in very large condition numbers. Nevertheless, the condition number only depends on the data sites $\mathcal{A}$ and a great amount of techniques have been developed for solving this problem [6, 16]

All these properties can be analysed in the case of radial basis functions with the concept of **native space**. Given a radial basis function $\phi(r)$, the goal is to find a space generated by $\phi(r)$ and providing this space with the standard properties of Hilbert spaces. The first step is to define the function space $H_\Phi(\Omega)$ of linear combinations

$$H_\Phi(\Omega) = span \{\mathbf{x} \mapsto \phi(\| \mathbf{x} - a_i \|) : \mathbf{x} \in \Omega\}$$

and then, to define an inner product in such a manner that the **native space** $\mathcal{N}_\Phi(\Omega)$ is the completion of $H_\Phi(\Omega)$ with respect to the induced norm. There are a number of technical details concerned with this construction which are discussed in the papers [55, 56] by Schaback or [72] by Wendland. In the case of strictly positive definite functions the native space can be characterized in terms of Fourier transforms and every $f \in \mathcal{N}_\Phi(\Omega)$ can be recovered from $\hat{f}$. These native spaces can be reinterpreted as generalizations of the standard Sobolev spaces.

Native spaces for strictly conditionally positive definite functions can also be constructed with more technical details [56]. For example the native spaces for thin plate spline and biharmonics can be shown to be

$$BL_k(\mathbb{R}^n) = \{u \in C(\mathbb{R}^n) : \partial^\alpha u \in L_2(\mathbb{R}^n), \forall \, |\alpha| = k\},$$

called Beppo-Levi spaces of order k. These were the spaces studied by Duchon in his classical papers about surface splines.

It is known [16] that the native space of Gaussians contains the so-called bandlimited functions, i.e., functions whose Fourier transform is compactly supported. This suggests the multivariable version of Shannon's theorem. We now end with the n-dimensional version of the theorem that opened the book [34]

Theorem 4. (Shannon's in $\mathbb{R}^n$) Suppose $f \in C(\mathbb{R}^n) \cap L_1(\mathbb{R}^n)$ such that its Fourier transform vanishes outside the cube $Q = \left[-\dfrac{1}{2}, \dfrac{1}{2}\right]^n$. Then f can be uniquely reconstructed from its values on $\mathbb{Z}^n$, that is,

$$f(\mathbf{x}) = \sum_{\xi \in \mathbb{Z}^n} f(\xi) \operatorname{sinc}(\mathbf{x} - \xi)$$

$$\mathbf{x} = (x_1, x_2, \ldots, x_n) \in \mathbb{R}^n, \operatorname{sinc} \mathbf{x} = \prod_{i=1}^n \frac{\sin(\pi x_i)}{\pi x_i}$$

Native spaces have shown to be an appropriate setting for studying error estimates of radial basis functions approximation. The subject have been widely studied and appears in the pioneering work of Duchon [12] and Madych and Nelson [41, 42]. Nevertheless there are many open problems, not only on this respect but in all about radial basis functions. Currently radial basis functions are being applied not only in interpolation problems, but also in diverse fields as computer graphics, computational fluid dynamics or numerical solution of partial differential equations, in particular Schrodinger's; a problem of great importance in modern technology.

8.6   Matlab codes for Chap. 8

```
function graphmultiq(L,x,y,a,b,h,e)
% x,y column vectors for data points
[x1,y1]=meshgrid(a:h:b);
[m,n]=size(x1);
z1=zeros(m,m);
for i=1:m
   for j=1:m
       z1(i,j)=feval('splmultiq',L,x,y,x1(i,j),y1(i,j),e);
   end
end
mesh(x1,y1,z1);

%--------------------------


function A = matmultiq(x,y,e)
% A :interpolation matrix for multiquadric
%x,y : centers coordinates   enter as column
%e:multiquadric parameter
N=length(x);
o = ones(1,length(x));

r = sqrt( (x*o -(x*o)' ).^2 + (y*o -(y*o)').^2 );

A = multiq(r,e);


%--------------------------------------------------
function p = multiq(r,e)

p=sqrt(1+(e*r).^2);
%--------------------------------------------------

function L =splinemultiq(x,y,z,e)
```

```
%L:vector with weights of radial basis function
%e:multiquadrics parameter
%x,y,z :coordinate vectors of the data points
%build interpolation matrix A
A=matmultiq(x,y,e);
%Solve the system A*L=z
L = A\z;
```

```
%------------------------------------------------
```

```
function S = splmultiq(L,x0,y0,x,y,e)
% x0,y0 data points in the plane
% (x,y) evalution point
N=length(x0);
x1=x*ones(N,1);
y1=y*ones(N,1);
S=L'*multiq(sqrt((x1-x0).^2+(y1-y0).^2),e);
```

```
%------------------------------------------------
```

the reader may run these programs in the
following way:

choose values for a,b,h,e.
Once we have the column vectors x,y,z
with the coordinates of the knots,
these files are run in the following order

```
>>L=splinemultiq(x,y,z,e)
>>graphmultiq(L,x,y,a,b,h,e)
```

References

1. Ahlberg J, Nilson HE, Walsh JL (1967) The theory of splines and their applications. Academic Press, New York
2. Aronszajn N (1950) Theory of reproducing kernels. Trans Amer Math Soc 68:337–404
3. Atteia M (1992) Hilbertian Kernels and Spline functions. Elsevier Science, North-Holland
4. Bertero M, Poggio T, Torre V (1986) Ill-posed problems in early vision. Artifitial intelligence Laboratory Memo, No. 924. MIT
5. Bouhamidi A, Le Méhauté A (1999) Multivariate interpolating (m, l, s)-splines. Adv Comput Math 11:287–314
6. Buhmann MD (2003) Radial basis functions. Cambridge University Press, Cambridge (Cambridge Monographs on Applied and Computational Mathematics)
7. Cheney EW, Light WA (2000) A course in approximation theory. Brooks Cole Publishing Company, Pacific Grove
8. Cheney W (2001) Analysis for applied mathematics. Springer
9. Choquet-Bruhat Y, DeWitt-Morette C, Dillard-Bleick M (1982) Analysis, manifolds and physics, rev. ed. North-Holland
10. Davis PJ (1975) Interpolation and approximation. Dover
11. Duchon J (1976) Interpolation des fonctions de deux variables suivant le principe de la exion des plaques minces, Rev. *Francaise Automat.* Informat Rech Oper Anal Numer 10:5–12
12. Duchon J (1977) Splines minimizing rotation-invariant semi-norms in Sobolev spaces. In: Schempp W, Zeller K (eds) Constructive theory of functions of several variables, Oberwolfach 1976. Springer Lecture Notes in Math, 571. Springer-Verlag, Berlin, pp 85–100
13. Duchon J (1978) Sur l'erreur d'interpolation des fonctions de plusieurs variables par les Dm-splines. Rev Francaise Automat Informat Rech Oper Anal Numer 12:325–334
14. Duchon J (1980) Fonctions splines homogenes a plusiers variables. Universite de Grenoble
15. Engl HW, Hanke M, Neubauer A (1996) Regularization of inverse problems. Springer
16. Fasshauer G (2007) Meshfree approximation methods with matlab. Interdisciplinary mathematical sciences. World Scientific Publishers
17. Franke R (1985) Thin plate splines with tension. Comput Aided Geom Des 2:87–95
18. Franke R (1982) Scattered data interpolation: tests of some methods. Math Comp 48:181–200
19. Friedlander FG, Joshi M (2008) Introduction to the theory of distributions. Cambridge University Press
20. Gel'fand IM, Shilov GE (1964/1968/1967). Generalized functions, Vol 1–3. Academic Press, New York
21. Gelfand IM, Fomin SV (1963) Calculus of variations. Prentice-Hall
22. Gibson JJ (1979) The ecological approach to visual perception. Houghton Mifflin
23. Girosi F, Jones M, Poggio T (1995) Regularization theory and neural networks architectures. Neural Comput 7(2):219–269
24. Goldstine H (1980) A history of the calculus of variations from the 17th through the 19th Century

H. Montegranario, J. Espinosa, *Variational Regularization of 3D Data,*
SpringerBriefs in Computer Science, DOI 10.1007/978-1-4939-0533-1,
© The Author(s) 2014

25. Golomb M, Weinberger HF (1959) Optimal approximation and error bounds. In: Langer RE (ed) Numerical approximation. University of Wisconsin Press, Madison, pp 117–190
26. Giaquinta M, Hildebrandt S (1996) Calculus of VariationsI:the lagrangianformalism. Springer-Verlag
27. Grimson WEL (1981) From images to surfaces: a Computational study of the human early vision system. MIT Press, Cambridge
28. Hadamard H (1923) Lectures on the Cauchy problem in linear partial differential equations. Yale University Press, New Haven
29. Hansen PC (1998) Rank-deficient and discrete ill-posed problems: numerical aspects of linear inversion. Society for Industrial Mathematics
30. Hardy RL (1971) Multiquadric equations of topography and other irregular surfaces. J Geophys Res 76:1905–1915
31. Hilbert D, Courant R (1989) Methods of mathematical physics. Wiley
32. Hormander L (1963) Linear partial differential operators. Springer-Verlag, Berlin
33. Kincaid D, Cheney W (2001) Numerical Analysis: mathematics of scientific computing, 3rd edn. Brooks Cole
34. Kybic J, Blu T, Unser M (Aug 2001) Generalized sampling: a variational approach. Part I-theory. IEEE Trans Signal Process 50:1965–1976
35. Lanczos C (1970) The variational principles of mechanics, 4th edn. University of Toronto Press
36. Laurent PJ (1972) Approximation et optimisation. Hermann, Paris
37. Light WA, Wayne H (1998) On power functions and error estimates for radial basis function interpolation. J Approx Theory 92:245–266
38. Light W (1998) Variational methods for interpolation, particularly by radial basis functions. University of Leicester. Technical Report
39. Light W, Wayne H (1997). Spaces of distributions and interpolation by translates of a basis function. University of Leicester. Technical Report
40. Luenberger D (1969) Optimization by vector space methods. Wiley
41. Madych WR, Nelson SA (1983) Multivariate interpolation: a variational theory, manuscript.
42. Madych W, Nelson S (1988) Multivariate interpolation and conditionally positive definite functions. Approx Theory Appl 4:77–89
43. Marr D (1982) Vision: a computational investigation into the human representation and processing of visual information. W. H. Freeman and Company, New York
44. Marr D, Poggio T (1976) Cooperative computation of stereo disparity. Science 194:283–287
45. Meinguet J (1979) An intrinsic approach to multivariate spline interpolation at arbitrary points. In: Sahney NB (ed) Polynomial and spline approximations. Reidel, Dordrecht, pp 163–190
46. Meinguet J (1979) Basic mathematical aspects of surface spline interpolation. In: Hammerlin G (ed) Numerische integration. Birkhauser, Basel, pp 211–220
47. Meinguet J (1984) Surface spline interpolation: basic theory and computational aspects. In: Singh SP, Burry JHW, Watson B (eds) Approximation theory and spline functions. Reidel, Dordrecht, pp 127–142
48. Micchelli CA (1986) Interpolation of scattered data: distance matrices and conditionally positive definite functions. Constr Approx 2:11–22
49. Montegranario H, Espinosa J (2007) Regularization approach for surface reconstruction from point clouds. Appl Math Comput 188:583–595
50. Morozov VA (1993) Regularization methods for ill-posed problems. CRC Press
51. O'Neill B (1966) Elementary differential geometry. Academic Press
52. Poggio T, Torre V (1984) Ill-posed problems and regularization analysis in early vision. MIT Artifitial Intelligence Lab. A.I. Memo 773
53. Raja V, Fernandes KJ (eds) (2008) Reverse engineering. An industrial perspective. Springer
54. Saxena A, Sahay B (2005) Computer aided engineering design. Springer
55. Schaback R (1999) Native Hilbert spaces for radial basis functions I. In: Muller MW, Buhmann MD, Mache DH, Felten M (eds) New developments in approximation theory. Birkhauser, Basel, pp 255–282

56. Schaback R (2000) A unified theory of radial basis functions. Native Hilbert spaces for radial basis functions II. J Comput Appl Math 121:165–177
57. Schoenberg IJ (1938) Metric spaces and completely monotone functions. Ann of Math 39:811–841
58. Schoenberg IJ (1964) Spline functions and the problem of graduation. Proc Nat Acad Sci 52:947–950
59. Schwartz L (1966) Théorie des distributions. Hermann, Paris
60. Schwartz L (1966) Mathematics for the physical sciences. Addison-Wesley
61. Stoker JJ (1969) Differential geometry. Wiley-Interscience
62. Terzopoulos D (1988) The computation of visible-surface representation. IEEE Trans Pattern Anal Mach Intell 10(4):417–438
63. Tikhonov N, Arsenin VY (1977) Solutions of ill-posed problems. Winston, Washington D.C
64. Timoshenko S, Woinowsky S (1981) Theory of plates and shells. McGraw-Hill
65. Timoshenko SP, Gere JM. (1952) Theory of elastic stability. Wiley
66. Unser M, Zerubia JA 1998 Generalized sampling theory without band-limiting constraints. IEEE Trans Circuits Syst-II: Analog And Digit Signal Processing 45(8, August 1998):959
67. Vinesh R, Fernández KJ (eds) (2008) Reverse engineering. An industrial perspective. Springer
68. Wahba G, Luo Z (1997) Smoothing spline ANOVA ts for very large, nearly regular data sets, with application to historical global climate data, in The heritage of P. L. Chebyshev: a Festschrift in honor of the 70th birthday of T. J. Rivlin. Ann Numer Math 4(1–4):579–597
69. Wahba G (1979) Convergence rate of "thin plate" smoothing splines when the data are noisy. Springer Lecture Notes in Math 757:233–245
70. Wahba G (1990) Spline Models for Observational Data. CBMS-NSF, Regional Conference Series in Applied Mathematics, SIAM, Philadelphia
71. Wahba G, Wendelberger J (1980) Some new mathematical methods for variational objective analysis using splines and cross validation. Mon Wea Rev 108:1122–1143
72. Wendland H (2010) Scattered data approximation. Cambridge University Press
73. Zeidler E (1984) Nonlinear functional analysis and its applications: part II and III. Springer
74. Zexiao X, Jianguo W, Qiumei Z (2005) Complete 3D measurement in reverse engineering using a multi-probe system. Int J Mach Tool Manu 45:1474–1486